Amitava Dasgupta

Effects of Herbal Supplements in Medicine

Patient Safety

Edited by
Oswald Sonntag and Mario Plebani

Volume 2

Dedicated to my wife Alice

Preface

Herbal supplements are available without prescription in many countries throughout the world and account for more than US$30 billion in sales. A majority of the US population (25–40%) use herbal supplements and alternative medicines are major forms of therapy in Third World countries, used by as much as 80% of the population. Contrary to the popular belief that herbal remedies are safe and effective, many herbal supplements have known toxicity and unexpected laboratory test results may be early indications of such toxicity. In addition, some herbal products such as St. John's wort can interact with many Western drugs, causing increased clearance of such drugs and hence treatment failure. This monograph provides information on how herbal supplements affect laboratory test results and thus patient safety. This book is a comprehensive and concise practical guide for laboratory professionals, physicians and other healthcare professionals. The emphasis is on providing clinically relevant information rather than discussing mechanisms of such effects in detail, although brief explanations are provided for some unexpected test results.

Amitava Dasgupta
Houston, Texas, USA

Contents

1 Commonly used herbal supplements

1.1 Introduction

Throughout history, herbal remedies were often the only medicines available and even today many Western medicines are derived from medicinal plants. For example, the antibiotic streptomycin is derived from a soil bacterium (*Streptomyces griseus*) and the immunosuppressant cyclosporine, which is used to prevent host versus graft disease in transplant recipients, is derived from a soil fungus. Several anticancer drugs such as vincristine, vinblastine and paclitaxel (Taxol) are also derived from plants and the cardioactive drug digoxin is extracted from foxglove plants (*Digitalis lantana*). If Western drugs are prepared from plants, an extensive manufacturing process is used to isolate the active natural ingredient from other crude and often toxic plant products. In addition, the amount of active ingredient is maintained nearly constant in the final product using rigorous quality control procedures. By contrast, herbal supplements are crude plant extracts and any quality control procedures applied in the preparation of such supplements are not as rigorous as those during preparation of Western drugs. In crude herbal preparations, the active ingredients may be present along with other toxic substances.

Western medicines may also cause drug toxicity but owing to rigorous research such toxic effects are known to healthcare providers. The toxicities of herbal supplements have not been well studied. For example, in the 1990s kava, a herbal supplement used as an anti-anxiety and sleep aid, was considered relatively safe and as effective as some prescription antidepressant drugs (tricyclic antidepressants). However, it is now well established that long-term use of kava may cause severe liver toxicity and even death due to liver failure. One argument that herbalists make is that these remedies have been used for thousands of years and native people did not experience toxicity. The truth is that in ancient times, native peoples chewed leaves or roots or prepared tea by boiling plant parts; thus, only small amounts of plant alkaloids were extracted and probably caused lesser toxicity. In addition, documentation of the toxicity of herbal remedies is poor because native people might not have spoken out as a mark of respect for the healers. Today, these supplements are prepared via modern extraction techniques using a combination of organic solvent and water and thus may contain more active ingredients as well as toxic substances present in the plant. Various types of herbal supplements available commercially in health food stores are listed in ▶Tab. 1.1.

1.2 Regulation of the sale of herbal supplements

The Food and Drug Administration of the United States Government (FDA) regulate all prescription and non-prescription drugs sold in the USA, but herbal remedies are sold according to the 1994 Dietary Supplement Health and Education Act and are

Tab. 1.1: Herbal preparation types

Herbal preparation	Method	Example
Water extract	Extract herb in hot water, yields water-soluble components	Herbal tea, chamomile tea
Tincture[a]	Extract with alcohol/water or water and another solvent	Echinacea, hawthorn, St. John's wort
Tablet/solid form[a]	The crude herb is dried and a certain amount is placed in a capsule, or the dried alcoholic extract of a herb is used	St. John's wort, kava, garlic, ginkgo

[a]Many herbal supplements are available as both a liquid extract and a solid capsule. Herbalists often believe that active ingredients are more stable in liquid extract than solid powder form.

classified as food supplements. In addition to herbal remedies, vitamins, minerals, amino acids, extracts and metabolites are all considered as food supplements. As long as manufacturers of herbal remedies do not claim any therapeutic benefit from use of a herbal supplement, they are not subject to FDA surveillance. However, the FDA requires that the following information must be present on the label of a herbal supplement.

General information

- Name of product (including the word 'supplement' or a statement that the product is a supplement)
- Net quantity of contents
- Name and place of business of manufacturer, packer, or distributor
- Directions for use

Supplement facts

- Serving size, list of dietary ingredients, amount per serving size (by weight), percent of daily value (%DV), if established
- If the dietary ingredient is a botanical, the scientific name of the plant or the common or usual name standardized in the literature; if the dietary ingredient is a proprietary blend (i.e., a blend exclusive to the manufacturer), the total weight of the blend and its components in order of predominance by weight

Other ingredients

- Nondietary ingredients, such as fillers, artificial colors, sweeteners, flavors or binders, listed by weight in descending order of predominance and by common name or proprietary blend

Consumers and healthcare providers can report suspected adverse effects to the FDA through the MedWatch program or through the Toxic Exposure Surveillance System of the American Association of Poison Control Centers. If there is sufficient evidence, the FDA will issue an alert. Since 2001, alerts have been issued by the FDA for various products

Tab. 1.2: Some dietary supplement alerts issued by the US FDA

Alert	Year	Comment
St. John's wort	2000	St. John's wort and all products containing it (herbal antidepressant) should be avoided if taking any medication owing to drug-herb interactions
Tiratricol	2000	Products containing tiratricol (triiodothyroacetic acid), a thyroid hormone analog, should be avoided
Aristolochic acid	2001	Found in some herbal weight loss products; a nephrotoxic substance
Comfrey	2001	Used as a tonic and for repairing bone; causes liver damage
Lipokinetix	2001	Weight loss product; causes liver damage
Ephedra	2002	Used in ma-huang, a popular weight loss product; very toxic and may even cause death
Kava	2001–2002	Herbal sedative/antidepressant; known hepatotoxic herb that may even lead to death
PC SPES and SPES	2002	Used to improve prostate health; should be avoided
Liqiang 4	2005	Produced by a Chinese company; contains the hypoglycemic agent glyburide
Red yeast rice	2007	Fermented rice used in traditional Chinese medicine as a blood purifier and tonic; consumers warned not to use this product for lowering cholesterol
Colloidal silver	2009	Indicated for immune boosting, but there is no scientific evidence

Source: FDA website at http://www.fda.gov/Food/DietarySupplements/Alerts/default.htm, accessed May 24, 2010.

(▶Tab. 1.2). When an adverse event is suspected, the question often arises as to how the contents can be determined or how the product can be tested. Unfortunately, such testing is beyond the scope of most clinical laboratories.

In other countries, herbal supplements are not well regulated. However, in Germany, herbal monographs called the German Commission E monographs are prepared by an interdisciplinary committee using historical information; chemical, pharmacological, clinical, and toxicological study findings; case reports; epidemiological data; and unpublished manufacturers' data. If there is an approved monograph for a herb, it can be marketed. Currently, there is a push for harmonization in the European market for herbal supplements. European Directive 2004/24/EC released in 2004 by the European Parliament and the Council of Europe now provides a basis for regulation of herbal supplements in European markets. According to this directive, herbal medicines released in the market require authorization by the national regulatory authority of each European country and the products must be safe. The safety of a supplement will be established based on published scientific literature, and when safety is not sufficient, this will be communicated to consumers. The Australian Government also created a Complementary Medicine Evaluation Committee in 1997 to address regulatory issues regarding herbal remedies. In Canada, the Federal Government implemented a policy on January 1, 2004 to regulate natural health products; naturopaths, traditional Chinese medicine practitioners, homeopaths and Western herbalists in Canada are concerned that this policy will eventually affect their access to the products they need to practice effectively.

1.3 Sale of herbal supplements and demographics of use

In the USA, sales of herbal remedies rocketed from $200 million in 1988 to over $3.3 billion in 1997. Within the European Community, sales of herbal remedies are also widespread, with estimated annual sales of US$7 billion in 2001 (1). Sale of herbal supplements was estimated to be $15.7 billion in 2000 and in 2003 it was increased to an estimated $18.8 billion (2). The popularity of herbal supplement use is steadily increasing among the general US population. According to a study by the National Center for Complementary and Alternative Health (NCCAM) of the National Institutes of Health (Department of Health and Human Services, United States Government), one-third of adults and one-tenth of children in the USA have used complementary and alternative medicine in the last month prior to the latest survey available. Typical adults using complementary and alternative medicines are females with a relatively high income and educational level. Children's use of complementary and alternative medicines is related to use of such supplements by their parents. Unfortunately, only half of adult patients disclose their use of herbal supplements to their clinicians (3). It is estimated that approximately 20,000 herbal products are available in the USA and in one survey, approximately 1 out of 5 adult reported using a herbal supplement within the past year. The ten most commonly used herbal supplements are echinacea (see color plate), ginseng (see color plate), ginkgo biloba (see color plate), garlic (see color plate), St. John's wort (see color plate), peppermint, ginger (see color plate), soy, chamomile and kava (4).

Although both people with chronic illness and healthy people use various herbal supplements, the former tend to use more herbal supplements than the latter. According to one report, 72% of people who use herbal supplements also use prescription medications and 84% of users of herbal remedies use over-the-counter drugs. Among herbs used by adults, the most common was echinacea (41%). The other common herbal products used by respondents in the study included ginseng (25%), ginkgo biloba (22%), garlic (20%), St. John's wort (12.2%), and ginger supplements (10.5%). Herbal supplements commonly used in the USA and indications for their use are listed in ▶Tab. 1.3. The most frequent condition for use of herbal supplements was head or chest cold (30%), followed by musculoskeletal conditions (16%) and stomach or intestinal conditions (11%). People who use herbal supplements are between 25 and 44 years of age, are uninsured or avoid medical treatments because they are too expensive, and in general are in poor health (5). Patients with HIV infection also widely use complementary and alternative medicines. According to one report, the alternative medicines most commonly used by these patients are vitamin C (63%), multivitamin and mineral supplements (54%), vitamin E (53%) and garlic (53%) (6).

Tab. 1.3: Herbal supplements commonly used in the USA[a]

Herbal supplement	Indication for use
Echinacea	Boosting immune system (prevention of colds and influenza)
Ginseng	Tonic for general wellbeing
Ginkgo biloba	Dementia, prevention of Alzheimer's disease
Garlic	Lowering cholesterol, prevention of heart disease
St. John's wort[b]	Herbal antidepressant
Ginger	Nausea and vomiting
Soy	Lowering cholesterol, blood pressure
Kava[c]	Anti-anxiolytic
Valerian	Sedative
Saw palmetto	Prostate conditions
Black cohosh	Symptoms of menopause
Ma-huang (ephedra)[c]	Weight loss
Licorice	Adrenal gland support
Milk thistle	Improving liver function
Comfrey[c]	Local use for broken bones
Dong quai	Symptoms of menopause
Hawthorn	Tonic for cardiac health
Chaparral[c]	Autoimmune disease and allergy

[a]Source: reference (4).
[b]Should not be taken without consulting a physician if taking any Western drug for treatment of a chronic condition because there are many clinically significant interactions between St. John's wort and various Western drugs.
[c]Very toxic herb and should be avoided.

1.4 Commonly used relatively safe herbal supplements

Although most of the commonly used herbal supplements are relatively safe, toxic herbal products such as kava, comfrey, chaparral and weight loss products such as ma-huang are also commonly used. The popular concept that anything natural is safe is wrong because severe toxicity and even death may result from ingestion of toxic herbal supplements. Unfortunately, owing to a lack of authority, the FDA is unable to ban sales of some of the very toxic products and despite issuing customer advisories, some of these supplements are readily available in health food stores. This chapter focuses on commonly used herbal supplements that are relatively safe. Hepatotoxic herbs are discussed in chapter 2 and moderately toxic and toxic herbs are listed in chapter 11.

Aloe vera

Aloe vera is a plant of the lily family that has been used medicinally for over 5000 years by Egyptians, Indians, Chinese and Europeans. Aloe vera gel is an aqueous extract of the leaf pulp and may contain more than 70 active compounds. Traditionally, aloe vera is used topically for wound healing and for treating various skin conditions and it has also been claimed that aloe vera gel has anti-inflammatory, antioxidant, anti-aging and immune-boosting properties. Aloe vera gel is found in many skin lotions, hand creams and skin screens, and is also used as a laxative. Scientific studies have indicated that aloe vera has wound healing properties but its safety for oral use has not been confirmed.

Black cohosh

Black cohosh, a plant native to North America, has been used by Native Indians to treat various disorders in women including menopausal symptoms. Clinical studies have shown that black cohosh is effective in relieving symptoms of menopause including hot flashes and mood swing. The German health authorities approved the use of black cohosh (Remifemin brand) at 40 mg/day for 6 months for menopausal symptoms. Black cohosh also has a good safety record if used for 6 months or less. However, there are few case reports of severe liver toxicity after use of black cohosh, although the cause of toxicity could not be solely attributed to black cohosh in all cases. Although older research papers report that black cohosh has an estrogen-like effect, recent findings show that it is not estrogenic. Black cohosh should not be used by pregnant women because it may cause miscarriage.

Cranberry

Native Americans used cranberry as a food, dye and medicine. Cranberry sauce and cranberry juice are commercially available and are used as food. However, cranberry juice is widely used for symptomatic relief of urinary tract infection (UTI). Cranberry juice is also given to people to help reduce urinary odor. It is also believed that cranberry juice can decrease the rate of formation of kidney stones. Other than sauce and juice, cranberry powders in capsules are also available as a dietary supplement. Approximately 60% of women experience UTI in their lifetime and many also suffer from recurring UTI. Cranberry juice is an effective treatment for symptomatic relief of UTI. In one study, the authors reviewed data from nine clinical trials and concluded that cranberry products significantly reduced the incidence of UTI over 12 months, particularly in women with recurrent infections (7). Cranberry extract may also have activity against *Helicobacter pylori* (responsible for the majority of peptic ulcers) and is effective in lowering total cholesterol and LDL cholesterol in diabetic patients (8).

Dong quai

Dong quai is the popular name for *Angelica sinensis*, which grows in China, Japan and Korea. The dried root of this plant has been used by herbalists for treatment of menopausal symptoms. Dong quai is also considered as a female tonic.

It is commonly used in the USA for relief of menopausal symptoms including hot flashes, skin flushing, perspiration and chills. However, dong quai used alone has very little effect in relieving symptoms of menopause and traditional Chinese herbalist never prescribe dong quai as a single supplement. In one clinical trial involving 71 postmenopausal women who used dong quai for 24 weeks, no beneficial effect of dong quai was observed in preventing symptoms of menopause. Dong quai does not contain any significant amount of phytoestrogen. However, dong quai in combination with other herbs may have beneficial effects in reducing symptoms of menopause (9). In a recently published article, the authors commented that one Chinese herbal preparation Dang Gui Buxue Tang, which contains both dong quai and astragalus, is effective in reducing mild hot flashes in postmenopausal women but not effective in reducing the frequency of moderate to severe hot flashes (10).

Echinacea

Echinacea, a genus that includes nine species, is a member of the daisy family. Three species are found in common herbal preparations, *Echinacea angustifolia*, *Echinacea pallida* and *Echinacea purpurea*. Native Americans considered this plant as a blood purifier. Today, echinacea is used mainly as an immune stimulant to increase resistance to colds, influenza and other infections and is one of the best-selling herbs in the United States. Fresh and freeze-dried herb and an alcoholic extract of the herb are all commercially available. The aerial parts of the plant and the root, fresh or dried, can also be used to prepare echinacea tea. One of the constituents of echinacea, arabinogalactan, may have immune-boosting capacity. Sullivan et al. reported that at the cellular level, echinacea extract stimulates macrophages to produce interleukins, tumor necrosis factor and nitrous oxide, which provides evidence regarding the immunostimulatory capability of echinacea extract. Moreover, the authors observed that oral administration of echinacea extract can reduce the bacterial burden from infection by *Listeria monocytogenes* (11). However, some clinical studies using human subjects did not observe any protective effect of echinacea against colds. The ability of an oral echinacea preparation to prolong the time of onset of an upper respiratory infection was compared to a placebo (dummy supplement) in 302 volunteers from an industrial plant and several military institutions. Although subjects felt significantly better after taking echinacea, rigorous statistical analysis did not show any significant benefit of echinacea in delaying the onset of upper respiratory tract infection (12).

Fenugreek

Fenugreek seed is used as a spice in many countries and paste made from fenugreek seeds was used in many ancient cultures to treat a variety of diseases, including aiding in childbirth and as a lactation stimulant. The ancient Chinese used fenugreek as a tonic. Fenugreek is also an important component of Indian Ayurvedic medicine. Fenugreek is recommended by herbalists as a dietary supplement to reduce blood sugar, reduce blood cholesterol, increase appetite and treat digestive problems. In a study of 20 male volunteers aged 20–30 years, Abdel-Barry et al. demonstrated that an aqueous extract of fenugreek leaves effectively reduced blood sugar in these normal volunteers (13). Patients with diabetes are at an increased risk of cardiovascular disease.

A clinical trial revealed that fenugreek seed powder (25 g/day orally) administered for 21 days reduced cholesterol in five diabetic patients (14). Fenugreek is a safe dietary supplement and despite reports of dizziness and diarrhea after use, there have been no reports of any serious adverse effects. Fenugreek may cause abnormally low glucose in diabetic patients taking medicines and should be used with caution in diabetic patients taking either insulin or oral hypoglycemic agents.

Feverfew

Feverfew (*Tanacetum parthenium*) is a short perennial bush that grows along fields and roadsides. The most common use of feverfew is for the prevention of migraine headache, but it is also used for relief from arthritis pain, treatment of fever, menstrual problems, asthma and dermatitis. The name also suggests that it is a fever reducer. Feverfew is available as the fresh leaf, dried powdered leaf, capsules and tablets, fluid extract and oral drops. The active components of feverfew are parthenolide and other sesquiterpene lactones. During a migraine episode, serotonin is released from platelets. So migraine headache can be treated by inhibiting serotonin release with serotonin antagonists. Research has indicated that parthenolide and sesquiterpene lactones present in feverfew can inhibit serotonin release by platelets. Johnson evaluated the efficacy of feverfew in 17 patients with a history of migraine for at least 2 years. The patients took feverfew (50 mg) or placebo. Patients in the placebo group experienced migraine more often than patients receiving feverfew (15). Feverfew is a relatively safe dietary supplement. Adverse effects include dizziness, heartburn, indigestion and bloating, as well as mouth ulceration. Discontinuation of feverfew may produce muscle and joint stiffness, rebound of migraine symptoms, anxiety and poor sleep patterns. Feverfew should be avoided in pregnancy. Feverfew also interacts with anticoagulants such as warfarin and aspirin.

Flaxseed

Flaxseed (linseed) is most commonly used as a laxative. It is also used to lower high cholesterol and to treat hot flashes and breast pain. Flaxseed oil is also indicated for arthritis. Flaxseed contains a soluble fiber similar to that found in oat bran and is a very effective laxative agent. However, owing to its high fiber content, flaxseed should be taken with plenty of water. Flaxseed contains alpha-linolenic acid, which may have health benefits. Flaxseed is a safe and effective dietary supplement. However, it should not be taken with oral medication because its fiber may bind drugs, thus reducing their absorption from the stomach.

Garlic

Garlic is promoted for lowering cholesterol and blood pressure and thus preventing heart attack and stroke. Garlic is also indicated for preventing cancer. Garlic has various sulfur-containing compounds derived from allicin, which is formed by the action of an enzyme when garlic is chopped. Allicin is responsible for the characteristic odor of garlic and is mostly destroyed during cooking. Clinical trials to determine the lipid-lowering effects of garlic had mixed results. Although some studies did not find any statistically significant effects of garlic in lowering serum cholesterol and triglycerides, other studies showed beneficial lipid-lowering effects. Three clinical trials found reductions of 6.1–11.5% in cholesterol in garlic-treated patients, mainly due to reductions in low-density lipoprotein cholesterol. Triglycerides levels were also decreased after garlic treatment (16). Garlic is a safe and cheap supplement and there are only isolated reports of allergic reactions and minor adverse reactions. Topically applied garlic may lead to allergic reaction in some individuals. Garlic increases the anticoagulant effect of warfarin and patients on warfarin therapy should not take garlic supplements. However, these patients can use garlic as a spice in cooking. The usual dose is a 300-mg supplement taken twice or three times a day.

Mixtures of chopped garlic and oil left at room temperature can result in fatal botulism food poisoning according to the FDA. *Clostridium botulinum* bacteria are dispersed throughout the environment but are not dangerous in the presence of oxygen. The spores produce a deadly toxin in anaerobic (no oxygen), low-acid conditions and cause botulism. Oil in which chopped garlic is stored represents such an environment. In February 1989, three cases of botulism were reported in individuals who had consumed chopped garlic in oil, used as a spread for garlic bread. Testing of leftover garlic in the oil revealed a high concentration of *C. botulinum* and toxins, and the mixture had low acidity (pH 5.7). Refrigeration of chopped garlic in oil is important to avoid such food poisoning (17).

Ginger

Ginger, like garlic, is a popular culinary and medicinal herb. The ancient Chinese used ginger as a flavoring agent and for treating nausea. The characteristic odor and flavor of ginger are due to its volatile essential oils. In the USA, ginger is promoted for relieving and preventing nausea caused by motion sickness, morning sickness and other etiologies. Ginger is commercially available as the dried powdered root, capsules, tea, oral solution and as a spice. Phillips et al. studied the efficacy of ginger in preventing postoperative nausea and vomiting in 120 patients. The subjects were given 1 g of ginger or placebo or 100 mg of metoclopramide, an antiemetic medication. The authors concluded that ginger was equally effective as metoclopramide in preventing postoperative nausea and vomiting (18). A clinical study has also indicated that ginger is more effective in preventing motion sickness than the drug dimenhydrinate. Like garlic, ginger is a safe herbal supplement with no serious adverse reactions reported.

Ginkgo biloba

Ginkgo biloba is prepared from dried leaves of the ginkgo tree by organic extraction (acetone/water). After the solvent is removed, the extract is dried and standardized. Most commercial dosage forms contain 40 mg of this extract. Ginkgo trees were brought from China to Europe. Ginkgo fruits and seeds have been used in China since ancient times. Ginkgo biloba is sold as a dietary supplement to improve blood flow in the brain and peripheral circulation, to sharpen memory and to treat vertigo. Ginkgo is one of the best-selling herbs in the USA and herbalists recommend ginkgo in older people to prevent age related dementia and Alzheimer's disease.

Although earlier studies showed beneficial effects of ginkgo in improving memory and treating dementia due to Alzheimer's disease, a more recent clinical trial showed no beneficial effect. A study funded by the National Center for Complementary and Alternative Medicine in the USA was conducted between 2000 and 2008 in five academic medical centers in the USA involving more than 3000 volunteers aged 75 years or older who took 120 mg of ginkgo twice a day and were followed for an average of 6 years. No beneficial effects of ginkgo in lowering the overall incidence of dementia and Alzheimer's disease in this elderly population were observed (19).

One commonly reported adverse effect of ginkgo is bleeding. One report described a 70-year-old man who presented with bleeding from the iris into the anterior chamber of the eye 1 week after beginning a self-prescribed regimen consisting of a concentrated ginkgo biloba extract at a dosage of 40 mg twice a day. His medical history included coronary artery bypass surgery performed 3 years previously. His only medication was 325 mg of aspirin daily. After the spontaneous bleeding episode, he continued taking aspirin but discontinued the ginkgo supplement. Over a 3-month follow-up period he had no further bleeding episodes. Interaction between ginkgo and aspirin was considered the cause of his eye hemorrhage (20). Therefore, concurrent use of ginkgo and aspirin, as well as other non-steroidal anti-inflammatory drugs, may cause bleeding because ginkgolide B in ginkgo extracts is a potent inhibitor of platelet activating factor. Ginkgo should also be avoided by patients on anticoagulant therapy with warfarin because ginkgo increases the action of warfarin and may cause bleeding.

Ginseng

Ginseng is a widely used herbal product in China, other Asian countries and the USA, where it is one of the most common herbal supplements taken by the general population. For thousands of years the Chinese have used ginseng as a tonic and in emergency medicine to revive dying patients. Ginseng means 'essence of man' in Chinese. Although the term ginseng is used loosely to define many different herbs, the term Asian ginseng, which is the most common form of ginseng sold in the USA, is prepared from the root of plant *Panax ginseng* that grows naturally in Manchuria. American ginseng is prepared from *Panax quinquefolius* and Siberian ginseng from the root of *Eleutherococcus senticosus*. The most common preparation is ginseng root. In the American market, ginseng is sold as a liquid extract, tablets or capsules. Dried ginseng root is also available in health food stores. Ginseng is promoted as a tonic capable of invigorating the user physically, mentally and sexually. Ginseng also has a calming effect. Asian ginseng contains saponins known as ginsenosides, whereas Siberian ginseng is devoid of any ginsenosides. Because of the subjective nature of feeling more energetic or calm, it is hard to conduct clinical trials to determine the efficacy of ginseng. In one clinical trial involving college students who took 100 mg of ginseng twice daily for 12 weeks, students who took ginseng performed mathematical calculations at a higher speed than those who did not take ginseng. Another study involving 625 volunteers concluded that ginseng improved the quality of life among its users (16). Asian ginseng also has a mild blood-sugar-lowering effect.

Asian ginseng is a fairly safe herbal supplement. One fatality from the use of ginseng was attributed to contamination of the ginseng product with ephedra (16). In 1979, the term ginseng abuse syndrome was coined as a result of a study on 133 people who took ginseng for 1 month. Most subjects experienced central nervous system stimulation, but higher doses led to confusion and depression. Fourteen patients experienced ginseng abuse syndrome, characterized by symptoms of hypertension, nervousness, sleeplessness, skin eruption and morning diarrhea (21). Nevertheless, ginseng is considered as a generally safe herb for human consumption. Ginseng should be avoided in pregnancy and by individuals on anticoagulation therapy with warfarin. Ginseng can reduce the efficacy of warfarin. Owing to its effect on blood sugar, diabetic patients taking medication for blood sugar control should consult their physician before taking ginseng.

Hawthorn

Hawthorn fruit has been used historically to treat heart disease, especially in European countries. Today, extracts of hawthorn leaves and flowers are indicated for treatment of heart failure and coronary artery disease. Both liquid extract and dried extracts in capsule form are available commercially. Clinical studies have indicated that there is a significant benefit in symptom control and physiological outcome from hawthorn extract as an adjunct treatment for chronic heart failure (22). Hawthorn is safe in most adults when used for a short period of time and the side effects are minimal.

Licorice root

Licorice root has been used in Indian Ayurvedic medicine, Asian medicine and in European countries for many centuries. Licorice extract has a sweet taste and is often used as a flavoring agent to mask the bitter taste of preparations and as an expectorant. Licorice root has been used as a dietary supplement for treating stomach ulcers, bronchitis and sore throats. It is also indicated for treating viral illness such as hepatitis C. Licorice is available as a liquid root extract or a dried powder in capsule form. Owing to its sweet taste, licorice is used as a flavoring agent in making candy. The main component is glycyrrhizic acid, although a variety of other active compounds such as triterpene saponins and flavonoids are found in licorice extracts. Clinical studies have reported beneficial effects of both licorice and glycyrrhizin consumption, including anti-ulcer and anti-viral effects and protection of the liver from injury. If consumed for a long time or in high amounts, licorice extract acid may inhibit the activity of the enzyme that inactivates cortisol. However, this effect is reversed when licorice is discontinued. Otherwise licorice is a safe and effective supplement (23). Use of licorice root extract for 4–6 weeks is safe and usually does not lead to any serious adverse effects. However, glycyrrhizin consumption may increase blood pressure and individuals with hypertension are advised not to take licorice supplements without consulting their physician. Licorice may also cause hypokalemia.

Milk thistle

For more than 2000 years, seeds of milk thistle have been used to treat liver disease. In the USA, milk thistle is mainly used to treat viral infection and cirrhosis of the liver. An active component of milk thistle, silymarin, protects the liver from a variety of toxins in an animal model (16). Clinical trials to evaluate the efficacy of milk thistle in treating liver cirrhosis reported mixed results. In one trial, milk thistle had beneficial effects in patients who had mild alcoholic liver cirrhosis. In another clinical trial, blood tests for liver damage showed improvement in liver cirrhosis patients who took milk thistle supplements. However, other studies reported no beneficial effect of milk thistle in treating liver cirrhosis. Milk thistle extracts are known to be safe and well tolerated and adverse reactions are usually minimal.

Saw palmetto

Saw palmetto is a dwarf palm tree that grows in the southern part of the USA. Native Americans used saw palmetto to treat genitourinary conditions. In the early 20th century it was used in conventional medicine as a mild diuretic and as a treatment for benign prostatic hypertrophy (BPH). The recommended dose of saw palmetto is 1–2 g of dried saw palmetto fruit or 320 mg of lipophilic extract. This herbal supplement is mostly taken to treat enlarged prostate. Carraro et al. performed a large double-blind study to evaluate the efficacy of saw palmetto in 1069 men with moderate BPH and observed no benefit (24). Adverse effects of saw palmetto seem to be minimal, with gastrointestinal upset the most common. However, patients on hormonal therapy for prostate cancer should not take saw palmetto without first consulting their physician because adverse drug-herb interactions may occur.

St. John's Wort

St. John's wort is prepared from a perennial aromatic shrub with bright yellow flowers that bloom from June to September. The flowers are believed to be most abundant and brightest around June 24, the day traditionally believed to be the birthday of John the Baptist. Therefore, the name St. John's Wort became popular for this herbal product. The flowering tops of St. John's wort and the leaves and stem are used to make extracts or dried powder for capsule products. St. John's wort is one of the best-selling herbal supplements in the USA and is mostly used for treating depression, anxiety or sleep disorder. Many chemicals have been isolated from St. John's wort, but hypericin and hyperforin are mainly responsible for the antidepressant effects. Most commercially available St. John's wort preparations are standardized to contain 0.3% hypericin. Linde et al. analyzed data from 23 clinical trials involving St. John's wort (1757 outpatients) and concluded that it is effective in treating mild to moderate depression. In addition, on the basis of data from eight of these studies, it was suggested that St. John's wort has equivalent efficacy to low-dose treatment with antidepressant medications (amitriptyline or imipramine). Interestingly, St. John's wort has fewer side effects than these antidepressants (25).

St. John's wort is fairly safe but may induce photosensitivity, especially in fair-skinned individuals, following exposure to sunlight or UV light. After taking St. John's wort for 4 weeks, a 35-year-old woman complained of stinging pain in sun-exposed areas.

Her symptoms improved 2 months after she discontinued use of the product (26). There are a few case reports describing episodes of hypomania (irritability, disinhibition, agitation, anger, insomnia and difficulty in concentrating) after using St. John's wort. The major problem in taking these supplements are the reported interactions with many Western drugs. St. John's wort induces liver enzymes, which metabolize many drugs, and thus reduces the effective blood level of many drugs. In addition, by other complex mechanism, St. John's wort also reduces blood levels of many drugs that are not primarily metabolized by the liver. Therefore, treatment failure may result in such patients if they take St. John's wort. Patients on warfarin therapy, transplant recipients and patients being treated for AIDS may face serious treatment failure if they take St. John's wort. Chapter 5 includes a more detailed discussion on drug interactions with St. John's wort.

Senna

Senna is used mainly as a laxative. In the USA it is available as leaf extract, fruit extract or as powder. When used at the recommended dose, senna laxative is safe and effective if taken for a short duration (<7 days). However, abuse of senna may cause problems because of loss of potassium from the body. Senna abuse may also cause abdominal pain, severe diarrhea and weight loss. There is a case report of acute liver failure in a 52-year-old woman who ingested 1 L of herbal tea each day for over 3 years using herbal tea bags containing 70 g of dry senna fruit. She developed acute liver failure and renal impairment requiring intensive care therapy (27).

Soy

Soy is available as a dietary supplement in the form of tablets and capsules and may contain soy proteins, isoflavones or both. Soybeans can be cooked and eaten or can be used to make soy milk, tofu or other food products. Soy supplements are indicated for lowering cholesterol, for weight loss and for relieving symptoms of the menopause such as hot flashes and osteoporosis. It is believed that soy supplements can also prevent cancer and improve memory. Soy protein is considered as a complete protein because it contains ample amounts of all the essential amino acids and several other nutrients. Therefore, soy protein is considered equivalent to animal protein. Soy also contains isoflavones, a group of compounds with many beneficial effects. Several studies in humans and in animal models have indicated that consumption of soy protein reduces body weight and fat mass. In obese humans, soy supplements reduce body weight and fat mass, in addition to reducing cholesterol and triglycerides. Soy supplements reduce cholesterol via a novel mechanism involving activation of receptors in the liver that remove low-density lipoprotein cholesterol from the blood (28). Another study indicated that soy consumption reduces the incidence of hot flashes in menopausal women (29). However, the cancer prevention capacity of soy supplements has not been demonstrated by rigorous research. Soy is an effective and safe supplement with only a few minor side effects reported. Some people may experience nausea, bloating or constipation following consumption of soy supplements. Allergic reaction and rash can occur in rare cases.

Turmeric

Turmeric is used as a spice in many Asian countries and is an important component of traditional Chinese medicine and Indian Ayurvedic medicine. The finger-like underground stems (rhizomes) of turmeric are dried and used as a spice or taken as a powder in capsule form. Liquid extract of turmeric is also commercially available. Turmeric can also be used as a paste for topical application. The most active component of turmeric is curcumin. Scientific research has shown that turmeric has antibacterial, antiviral, antifungal, antioxidant and anticancer activities and thus has potential against various malignant diseases, including arthritis, Alzheimer's disease and other chronic illnesses (30). Turmeric is a safe and effective herbal supplement. High doses and long-term use may cause indigestion. Individuals with gall bladder disease should not use turmeric supplements.

Valerian

Valerian is a perennial herb that grows in North America, Europe and Western Asia. The crude valerian root, rhizome or stolon is dried and is used as is or as an extract. Valerian is available as a capsule, liquid extract or tea and is used as a sleeping aid. Over 40 compounds have been isolated from valerian root, but valepotriates are probably responsible for the sedative activity. Valerenic acid, another component of valerian, may also have a pharmacological action similar to that of pentobarbital (a sedative drug). Leathwood et al. conducted a study in 128 volunteers and concluded that valerian significantly improved subjective sleep quality in habitually poor or irregular sleepers (31). A more recent study demonstrated that a combination of valerian and hops is an effective sleeping aid. Valerian supports readiness to go to sleep while hops have a melatonin-like effect. The authors concluded that the efficacy of the valerian+hops combination in sleep disorder can be explained from a scientific point of view (32). Studies have indicated that valerian is generally safe if used for short-term relief of insomnia (4–6 weeks). There are no reports on the long-term safety of valerian. Side effects are mild and include dizziness, upset stomach, headache and sleepiness in the morning.

1.5 Commonly used toxic herbs

Despite substantial medical reports of serious toxicity and FDA alerts, several toxic herbs such as kava, germander, chaparral, comfrey, pennyroyal and ma-huang are used by the general population for various reasons, but most commonly due to the belief that any natural supplement is safe. Monographs written by self-claimed experts and available in various herbal stores only focus on efficacy claims for herbal supplements, including toxic supplements, which are not supported by medical research. Moderately toxic and toxic herbs are discussed in Chapter 12. ▶Tab. 1.4 lists the most commonly used herbal supplements and toxicity values.

Tab. 1.4: Toxic herbs commonly used in the USA

Toxic herbal supplement	Organ-specific toxicity
Kava	Liver toxicity owing to the presence of kava lactones; death may result from prolonged use of this product
Ma-huang	Ephedra-containing herbal weight loss product that has been linked to several deaths
Chinese weight loss herbs	May be contaminated with aristolochic acid, a nephrotoxic agent
Comfrey	Hepatotoxic herb that must be avoided
Chaparral	Hepatotoxic herb that must be avoided
Yohimbine	Hallucinogen and may cause rapid heartbeat
Various Chinese herbs	Other than specific toxicity, may be contaminated with heavy metals
Indian Ayurvedic medicine	Other than specific toxicity, may be contaminated with heavy metals

1.6 Conclusions

One-third to half of all Western medicines are derived from plants and there is a strong possibility of finding new effective therapies from the natural environment. Certain herbal supplements are efficacious, but severe toxicities have also been reported following the use of herbal supplements. Although herbal supplements prepared by American manufacturers are usually free from heavy metals such as lead, arsenic and mercury, many herbal supplements imported from Asia, including traditional Chinese and Indian Ayurvedic medicines, may have high heavy metal content and use of such supplements for a prolonged period may cause severe heavy metal toxicity requiring hospitalization. By contrast, non-invasive complementary and alternative therapies such as massage therapy, energy healing, Reiki, other biofield therapies, yoga, meditation and prayer are safe. In addition, invasive therapy such as acupuncture is also relatively safe if practiced by an experienced professional. For those who would like to use herbal supplements, there are many safe supplements and this book provides information on some of those available. Doctors, nurses, pharmacists and other healthcare professionals also have a wealth of knowledge on complementary and alternative therapies and may help in identifying a suitable practitioner. Homeopathic remedies are also safe and even if they do not do any good, it is most likely they will not do any harm.

References

1. Mahady, G.B. (2001) Global harmonization of herbal health claims. J. Nutr. 131:1120S–1123S.
2. Kelly, J.P., Kaufman, D.W., Kelley, K., Rosenberg, L., Anderson, T.E, Mitchell, A.A. (2005) Recent trends in use of herbal and other natural products. Arch. Intern. Med. 165:281–286.

3. Lovett, E., Ganta, N. (2010) Advising patients about herbs and nutraceuticals: tips for primary care providers. Prim. Care 37:13–30.
4. Bent, S. (2008) Herbal medicine in the United States: review of efficacy, safety and regulation. J. Gen. Intern. Med. 23:854–859.
5. Gardiner, P., Graham, R., Legedza, A., Ahn, A.C., Eisenberg, D.M., Phillips, R.S. (2007) Factors associated with herbal therapy use by adults in the United States. Altern. Ther. Health Med. 13:22–28.
6. Standish, L.J., Greene, K.B., Bain, S., Reeves, C., Sanders, F., Wines, R.C. et al. (2001) Alternative medicine use in HIV positive men and women: demographics, utilization patterns and health status. AIDS Care 13:197–208.
7. Jepson, R.G., Craig, J.C. (2007) A systematic review of the evidence for cranberries and blueberries in UTI prevention. Mol. Nutr. Food Res. 51:738–745.
8. Lee, I.T., Chan, Y.C., Lin, C.W., Lee, W.J., Sheu, W.H.-H. (2008) Effect of cranberry extract on lipid profiles in subjects with type 2 diabetes. Diabet. Med. 25:1473–1477.
9. Anonymous (2004) Monograph. Angelica sinensis. Altern. Med. Rev. 9:429–433.
10. Hanes, C.J., Lam, P.M., Chung, T.K., Cheng, F.K., Leung, P.C. (2008) A randomized double blind placebo controlled study of effects of a Chinese herbal preparation (Dang Gui Buxue Tang) on menopausal symptoms in Hong Kong Chinese women. Climateric 11:244–251.
11. Sullivan, A.M., Laba, J.G., Moore, J.A., Lee, T.D. (2008) Echinacea induced macrophage activation. Immunopharmacol. Immunotoxicol. 30:553–574.
12. Melchart, W.E., Linde, K., Brandmaier, R., Lersch, C. (1998) Echinacea root extracts for the prevention of upper respiratory track infection – a double blind, placebo controlled randomized trial. Arch. Fam. Med. 7:541–545.
13. Abdel-Barry, J.A., Abdel-Hassan, I.A., Jawad, A.M., al Hakiem, M.H. (2000) Hypoglycemic effect of aqueous extract of leaves of Trigonella foenum in healthy volunteers. East. Mediterr. Health J 6:83–88.
14. Basch, E., Ulbricht, C., Kuo, C., Szapary, P., Smith, M. (2003) Therapeutic applications of fenugreek. Altern. Med. Rev. 8:20–27.
15. Johnson, E.S. (1985) Efficacy of feverfew as a prophylactic treatment of migraine. Br. Med. J. 291:569–573.
16. O'Hara, M., Kiefer, D., Farrell, K., Kemper, K. (1998) A review of 12 commonly used medicinal herbs. Arch. Fam. Med. 7:523–536.
17. Morse, D.L., Pickard, L.K., Guzewich, J.J., Devine, B.D., Dhayegani, M. (1990) Garlic-in oil associated botulism: episode leads to product modification. Am. J. Public Health 80:1272–1273.
18. Phillips, S., Ruggier, R., Hutchinson, S.E. (1993) Zingiber officinale (ginger) – an antiemetic for day care surgery. Anesthesia 48:715–717.
19. DeKosky, S.T., Williamson, J.D., Fitzpatrick, A.L., Kronmal, R.A., Ives, D.G., Saxton, J.A. et al. (2008) Ginkgo biloba for prevention of dementia: a randomized controlled trial. J. Am. Med. Assoc. 300:2253–2262.
20. Rosenblatt, M., Mindel, J. (1997) Spontaneous hyphema associated with ingestion of Ginkgo biloba extract. N. Engl. J. Med. 336:1108.
21. Siegal, R. (1979) Ginseng abuse syndrome: problems with panacea. J. Am. Med. Assoc. 241:1614–1615.
22. Pittler, M.H., Guo, R., Ernst, E. (2008) Hawthorn extract for treating chronic heart failure. Cochrane Database Syst. Rev. 23:CD005312.
23. Isbrucker, R.A., Buedock, G.A. (2006) Risk and assessment on the consumption of licorice root (Glycyrrhiza sp.), its extract and powder as a food ingredient with emphasis on the pharmacological and toxicology of glycyrrhizin. Regul. Toxicol. Pharmacol. 46:167–192.
24. Carraro, J.C., Raynaud, J.P., Koch, G. (1996) Comparison of phytotherapy (Permixon) with finasteride in the treatment of benign prostate hyperplasia: a randomized international study of 1,098 patients. Prostate 29:231–240.

25. Linde, K., Ramirez, G., Mulrow, C.D., Pauls, A., Weidenhammer, W., Melchart, D. (1996) St John's wort for depression: an overview and meta analysis of randomized clinical trials. Br. Med. J. 313:253–258.
26. Bove, G.M. (1998) Acute neuropathy after exposure to sun in a patient treated with St. John's wort. Lancet 352:1121–1122.
27. Vanderperren, B., Rizzo, M., Angenot, L., Haufroid, V., Jadoul, M., Hantson, P. (1995) Acute liver failure with renal impairment related to the abuse of senna anthraquinone glycoside. Ann. Phramacother. 39:1353–1357.
28. Sirtori, C.R., Lovati, M.R. (2001) Soy proteins and cardiovascular disease. Curr. Atheroscler. Rep. 3:47–53.
29. Kurzer, M.S. (2008) Soy consumption for reduction of menopausal symptoms. Inflammopharmacology 16:227–229.
30. Aggarwal, B.B., Sundrram, C., Malani, N., Ichikawa, H. (2007) Curcumin: the Indian solid gold. Adv. Exp. Med. Biol. 595:1–75.
31. Leathwood, P.D., Chauffard, F., Heck, F., Monoz-Boz, R. (1982) Aqueous extract of valerian root (Valeriana officinalis L.) improves sleep quality in man. Pharmacol. Biochem. Behav. 17:65–71.
32. Brattstrom, A. (2007) Scientific evidence for a fixed extract combination (Ze01019) from valerian and hops traditionally used as a sleep inducing aid. Wien. Med. Wochenschr. 157:367–370.

2 Abnormal liver function tests due to use of herbal supplements

2.1 Introduction

The liver is an essential organ for removal of toxic compounds from the body through metabolism. Many potentially toxic compounds enter the body through the gastrointestinal tract and are immediately transported via the portal vein to the liver and are removed by first-pass metabolism. Drug-induced liver damage has been well documented in the literature. Even safe analgesics such as acetaminophen (paracetamol) if taken in excess can cause serious liver toxicity and even death. Herbal supplements, like Western drugs, contain bioactive compounds that may cause liver toxicity and it has been well documented in the medical literature that many herbal supplements cause liver damage via various complex mechanisms. However, the greatest number of hepatotoxicity cases due to the use of herbal supplements has been documented by the Japanese Health Ministry: a significant number of cases of severe liver toxicity primarily involving women (ages 23–63 years) were reported following the use of two Asian herbal weight-loss products (1). A list of herbal supplements that may cause liver toxicity is provided in ▶Tab. 2.1.

2.2 Abnormal liver function tests and herbal supplements

The analytes used to reflect liver function (liver function tests or LFT's) include bilirubin, proteins, various enzymes found primarily in hepatocytes, and coagulation factors. Measurements of the serum or plasma activity of the enzymes aspartate aminotransferase (AST), alanine aminotransferase (ALT), γ-glutamyl transferase (GGT), and alkaline

Tab. 2.1: Herbal supplements and traditional Chinese medicines that may cause liver damage

Herbal supplement	Active component causing liver damage
Kava	Kava lactones
Chaparral	Nordihydroguaiaretic acid
Comfrey	Pyrrolizidine alkaloids
Germander	Teucrin A
Lipokinetix	Sodium usniate
Pennyroyal	Pulegone
Mistletoe	Viscotoxins, lectins
Chinese weight loss products	Ephedrine, *N*-nitrosofenfluramine

phosphatase (ALP) are routinely performed to investigate liver injury. These intracellular enzymes are released following hepatocyte death, and GGT and ALP also reflect injury to biliary cells. Total bilirubin and its conjugated and unconjugated forms are also useful biochemical parameter for detecting liver disease. Lactate dehydrogenase (LDH), although not liver-specific, may be elevated due to liver damage. Bilirubin is a breakdown product of erythrocytes (heme) and the liver is responsible for clearing bilirubin. Hepatocytes are responsible for conjugation of bilirubin for excretion into the bile. Total bilirubin is increased in jaundice, but can also be elevated for a variety of other reasons, including increased bilirubin production due to hemolytic anemia or internal hemorrhage, problems with bilirubin uptake by hepatocytes (viral hepatitis, cirrhosis) and post-hepatic obstruction of the biliary duct. The most common cause of increased unconjugated bilirubin is physiological jaundice observed in newborns due to premature breakdown of erythrocytes. In hepatobiliary diseases, both conjugated and unconjugated bilirubin concentrations are increased and result in elevated concentrations of total bilirubin. In herbal-induced liver damage, total bilirubin concentrations are often elevated. Normal values of the biochemical parameters used to investigate liver function tests and expected changes due to liver damage induced by herbal supplements are listed in ▶Tab. 2.2.

Various herbal products such as kava, chaparral, comfrey, pennyroyal, mistletoe and green tea extract can cause liver damage. Despite known toxicity and warnings issued by the authorities, these herbal supplements are still available in health food stores because herbal supplements are sold in the USA under the Dietary Supplement Act of 1994, under which, as long as no medical claim is made, a herbal supplement can be sold as a dietary supplement. When the US Food and Drug Administration (FDA) banned

Tab. 2.2: Reference interval for liver function tests and elevation due to hepatotoxicity caused by herbal supplements

Analyte	Reference interval		Comments
	Conventional unit	SI unit	
ALT	5–40 U/L	5–40 IU/L	ALT and AST are usually the most remarkably elevated liver enzymes
AST	10–40 U/L	10–40 IU/L	due to herbal-induced liver damage; elevation is usually ten- to 20-fold. However, a 60- to 70-fold increase in a case of kava-induced liver damage has been reported
ALP	30–120 U/L	30–120 IU/L	Elevated, usually three- to tenfold
GGT	0–51 U/L	0–51 IU/L	May be elevated, usually three- to fivefold
LDH	45–90 U/L	45–90 IU/L	May be elevated two- to fivefold
Total bilirubin	0.2–1.5 mg/dL	3.4–25.6 μmol/L	Elevation may be from two- to over tenfold, depending on the severity of liver damage
Conjugated bilirubin	0.1–0.5 mg/dL	1.7–8.6 μmol/L	May be elevated
Albumin	3.5–5 g/dL	35–50 g/L	May be normal or lower than normal

the use of phenylpropanolamine, all over-the-counter cold medications containing phenylpropanolamine were removed from all stores within 1 week. By contrast, despite FDA caution against the use of kava in 2003, this product is freely available in health food stores in Houston, where the author lives.

2.2.1 Kava

Kava, a herbal sedative with anti-anxiety or calming effects, is prepared by extracting rhizomes of *Piper methysticum,* a South Pacific plant. There are at least 72 different cultivars of this species, which differ both in appearance and in chemical composition, but the active component, kavalactones, are found in various amounts in all kava plants, mostly in rhizomes (2). Inhabitants of the South Pacific Islands prepare a kava-based drink by mixing fresh or dried rhizomes with cold water or coconut milk. Of the more than 18 kavalactones characterized, six are considered the primary constituents of kava extracts: kawain, dihydrokawain, methysticine, dehydromethysticine, yangonin, and desmethoxyyangonin (3). Scientific research conducted before 1988 indicated that kava was as efficacious as antidepressant drugs and tranquilizers in treating anxiety disorders, but by 2003, 11 cases of hepatic failure were reported. Seven required liver transplantation and there were four deaths. In January 2003, kava was banned in the European Union and Canada and the FDA issued another warning. To date, more than 100 cases of hepatotoxicity have been linked to kava exposure. Many have followed co-ingestion with alcohol, which seems to potentiate hepatotoxicity (4). Hepatotoxicity associated with kava consumption will be reflected by increases in serum activities of ALT, AST and GGT.

A 50-year-old male presented with jaundice and complained of fatigue, tanned skin and dark urine. His medical history was unremarkable and the patient was not on any medication and consumed no alcohol. However, he had been taking three to four capsules of kava extract daily for 2 months. His liver function tests showed that both AST and ALT were increased 60- to 70-fold above the reference intervals, ALP was 430 U/L, GGT was 691 U/L, LDH was 1132 U/L, total bilirubin was 16.3 mg/dL or 279.2 µmol/L and conjugated bilirubin was 12.4 mg/dL or 212.3 µmol/L (reference intervals see ▶Tab. 2.2). The patient was admitted to hospital and untrasonography showed a slight increase in liver size, but no ascites or portal vein thrombosis. Blood tests for hepatitis (A, B and C), HIV, cytomegalovirus and Epstein Barr virus were negative. The patient's condition following admission deteriorated over 48 h and he developed stage IV encephalopathy requiring intubation. The patient received a liver transplant 2 days later and recovered uneventfully. Biopsy of his damaged liver confirmed kava-induced failure; histological studies revealed extensive and severe hepatocellular narcosis and extensive lobular and portal infiltration of lymphocytes and numerous eosinophils (5).

The 18th case of kava-induced liver failure in Europe was documented in 2001 when a 60-year old woman was admitted to hospital with jaundice, fatigue, weight loss over several months, and icterus. Biochemical testing revealed elevated parameters, including increased total and conjugated bilirubin (30 mg/dL or 513.0 µmol/L), AST (921 U/L) and ALT (1350 U/L). Prothrombin time was <10%. Serological tests ruled out hepatitis and autoimmune causes of liver damage. Liver histology demonstrated extensive hepatocellular necrosis with intrahepatic cholestasis. The diagnosis was kava-induced liver failure. Because of her deteriorating condition, she received an orthotopic liver transplant 11 days after admission (6).

The dangers of using kava extracts are well recognized in the medical community, but kava has been used for thousands of years by native inhabitants of the South Pacific without experiencing serious hepatotoxicity. Traditionally, Pacific Islanders prepare kava by maceration of the rhizome in water, coconut milk or a combination of both. By contrast, commercial products are prepared as either ethanol- or acetone-based extracts. Whitton et al. found that traditional extracts of kava using water contained approximately 3% kavalactones, compared to 25% kavalactones when kava roots were extracted using 25% ethanol/75% water for commercial preparations. Dried extract prepared using 100% ethanol contained 100% kavalactones (7). Thus, commercial products provide a higher dose of kavalactones compared to the traditional preparation method. Unfortunately, the liver is unprepared for the higher concentrations of kava lactones afforded by commercial products and hepatic glutathione stores, which are key to kavalactone detoxification, are rapidly depleted. This mechanism can be compared to that for acetaminophen: toxicity is encountered once glutathione stores are depleted, allowing for the formation of a toxic metabolite. In fact, glutathione present in kava roots is water-soluble and the traditional preparation method has the added benefit of extracting small amounts of this compound as well. Glutathione is insoluble in ethanol and is absent in kava products prepared using 100% ethanol. However, when Teschke et al. compared the hepatotoxicity of aqueous, ethanolic and acetonic kava extracts and kava herb mixtures, adverse effects were observed for all preparations. Dose is a major contributing factor and some individuals may be more susceptible to kava effects because of genetic variations in drug-metabolizing enzymes (8).

2.2.2. Chaparral

Chaparral is a plant found in Southwestern USA and Northern Mexico. Leaves of this plant are used as a herbal therapy in the treatment of a wide variety of symptoms ranging from cold sores to muscle pain. It is also used for its purported antioxidant, anti-HIV and anti-cancer effects. It can be taken as a dried extract in capsule or tablet form. The leaves, stems and bark in bulk can also be used to brew tea. The active ingredient of chaparral is nordihydroguaiaretic acid (NDGA), a compound with proven anti-inflammatory and antioxidant properties. Unfortunately, chaparral may have severe hepatotoxicity, as documented by several reports of chaparral-associated hepatitis.

A 45-year-old woman presented to her physician with painless jaundice, anorexia, fatigue, pruritus, nausea and vomiting after taking 160 mg of chaparral daily for 10 weeks. Her significantly elevated liver function tests included: ALT 1611 U/L; AST 957 U/L; ALP 265 U/L; GGT 993 U/L; and bilirubin 11.6 mg/dL (198.4 µmol/L) (reference intervals see ▶Tab. 2.2). Hepatitis, cytomegaly and Epstein Barr virus were ruled out on the basis of blood test results. Liver biopsy showed acute inflammation with neutrophil and lympho-plasmacytic infiltration, hepatic disarray, and necrosis. The diagnosis was drug-induced cholestatic hepatitis following the use of chaparral (9).

Sheikh et al. reviewed 18 cases of suspected chaparral toxicity reported to the FDA and confirmed 13 cases of hepatotoxicity related to the herb. Clinical presentation included significant elevation of liver enzyme activities and other biochemical markers of hepatic injury 3–52 weeks after chaparral ingestion. In most cases, liver function tests returned to the reference interval after cessation of use, but at least two cases of fulminant hepatitis suggest that irreversible injury can occur (10).

2.2.3 Comfrey

Comfrey is a hardy perennial plant whose leaves and roots are used traditionally for wound healing, repairing broken bones, and in the treatment of arthritis, gout and psoriasis. To date there is no scientific evidence to support these claims. Comfrey contains pyrrolizidine alkaloids, which are well-known hepatotoxins. Russian comfrey is even more toxic than the European and Asian varieties because it contains more potent alkaloids. Endemic outbreaks in Jamaica, India and Afghanistan have been reported when cereals have become contaminated with comfrey or following the ingestion of comfrey tea (also known as bush tea). The mechanism of hepatotoxicity involves the formation of toxic metabolites from pyrrolizidine alkaloids present in comfrey. Use of comfrey is banned in Germany and Canada.

Yeong et al. described a case involving a 23-year-old male who presented with severe veno-occlusive disease and hypertension, and subsequently died from liver failure. Light microscopy and hepatic angiography demonstrated occlusion of the sublobular veins and small venous radicles of the liver associated with widespread hemorrhagic necrosis of hepatocytes. The patient was a vegetarian who had used comfrey leaves as a dietary supplement (11). Pyrrolizidine alkaloids and their *N*-oxides are found in comfrey, as well as in a variety of related herbs. Concentrations of two major pyrrolizidine alkaloids, symphytine and echimidine, vary widely between comfrey teas prepared from leaves purchased from different vendors.

2.2.4 Germander

Germander, *Teucrium chamaedrys*, is an aromatic plant in the mint family. Its blossoms are used as a folk medicine to treat dyspepsia, diabetes and gout. More recently, germander has also been used for weight loss. Although considered as safe in the past, the French authority banned this product in 1992 in response to 26 cases of hepatotoxicity following germander use, primarily by women, as a diet aid (12). Hyperbilirubinemia and markedly elevated ALT and AST were observed within 2 months of beginning the diet regimen. Laliberte and Villeneuve reported two cases: the first involved a 55-year-old woman who became ill after taking 1600 mg/day of germander for 6 months in an effort to reduce her cholesterol. She was admitted to hospital after a 3-week history of jaundice, asthenia, nausea and vomiting. She had no prior history of liver disease, blood transfusion, alcohol or illicit drug abuse. Her laboratory findings included total bilirubin 13.9 mg/dL (237.7 µmol/L); conjugated bilirubin 8.4 mg/dL (143.6 µmol/L), AST 1180 U/L; ALT 1500 U/L; and ALP 164 U/L (reference intervals see ▶Tab. 2.2). Her prothrombin time was within normal limits and serological tests for hepatitis (A, B and C) were negative. A liver biopsy sample showed bridging necrosis and collapse, lobular and portal tract inflammatory infiltration with polymorphonuclear and mononuclear cells and mild portal-tract fibrosis. Her hepatitis resolved within 2 months of discontinuing germander use. In the second case, a 45-year-old woman took germander (260 mg/day) for weight loss. Her laboratory results also suggested hepatotoxicity: total bilirubin

3.5 mg/dL (59.9 µmol/L); conjugated bilirubin 2.2 mg/dL (37.6 µmol/L); AST 417 U/L; ALT 451 U/L; and ALP 79 U/L (reference interval see ▶Tab. 2.2). She discontinued use of the herb and her health condition improved. After 4 months she felt much better and started taking germander again. Within 1 week, she again developed indications of liver toxicity: total bilirubin 17.0 mg/dL (289.0 µmol/L); conjugated bilirubin 13.2 mg/dL (225.7 µmol/L); AST 1245 U/L; ALT 784 U/L; and ALP 89 U/L (reference intervals see ▶Tab. 2.2). Hepatitis tests were again negative. The patient stopped using germander again and her liver function tests improved over a period of 3 months (13).

In general, the hepatotoxic effects of germander are observed within 9 weeks of use and are manifest as jaundice and elevated ALT and AST. After discontinuation of the herb, recovery may take 6 weeks to 6 months. The mechanism of toxicity is thought to be related to diterpenoid compounds that are metabolized to more potent toxins within the liver. The major hepatotoxic diterpene found in germander is teucrin A, which is bio-activated by cytochrome P-450 enzymes in the liver.

2.2.5 Oriental weight loss products and other oriental supplements

The most dangerous Oriental weight-loss product is ma-huang, which contains ephedra alkaloids. Ma-huang may cause severe cardiotoxicity, hepatotoxicity and even death. Ephedrine found in ma-huang and related weight loss products may be responsible for the toxic effects observed. As mentioned previously, Chaso and Onshido contain N-nitrosofenfluramine and should be avoided. Other Oriental herbal supplements containing multiple herbs and several Kampo medicines (notably Dai-Saiko-to and Sho-Saiko-to) are known to cause liver damage. Jin-Bu-Huan, a sedative analgesic, is also known to cause liver damage. Shou-Wu-Pian, a proprietary Chinese medicine used for treatment of dizziness, back pain and constipation, may also cause hepatitis. Shen-Min, another Chinese medicine, has also been associated with drug-induced hepatitis.

2.2.6 Lipokinetix

LipoKinetix has been promoted as a weight loss aid that increases metabolism. Therefore, no exercise is needed to burn excess fat in the body and this product is promoted as an easy method for weight loss instead of exercise. The product contains phenylpropanolamine, caffeine, yohimbine, diiodothyronine and sodium usniate. In 2002, seven patients who were using Lipokinetix developed acute drug-induced hepatitis and their viral hepatitis tests were all negative (14). Sodium usniate found in Lipokinetix is derived from usnic acid. Although sodium usniate has been used in skin creams and mouthwashes, liver toxicity following ingestion has been documented. Moreover, usnic acid is present in Kombucha tea (also known as Manchurian mushroom or Manchurian fungus tea), which is prepared by brewing Kombucha mushroom in sweet black tea. Acute liver damage due to drinking of this tea has been reported. There may be an inherent genetic susceptibility to usnic acid toxicity.

2.2.7 Pennyroyal

Pennyroyal, *Mentha pulegium*, is a plant of the mint genus whose leaves release a spearmint-like fragrance when crushed. Portions of the plant, as well as the essential oil, are used for a variety of purposes. Because of its strong fragrance, pennyroyal is often found as an additive in bath and aromatherapy products. Traditionally, pennyroyal has been brewed as a tea to be ingested in small amounts as an abortifacient and emmenagogue. The toxicity of pennyroyal has long been recognized and has been applied in its historical use for pest control. The plant and oil contain several components including pulegone. Pulegone is metabolized by the liver to a more toxic compound, methofuran, and is known to deplete hepatic glutathione stores (similar to the mechanism in acetaminophen overdose and probably for several of the herbs discussed above). A review of 18 cases of toxicity found that ingestion of as little as 10 mL of pennyroyal oil can cause severe toxicity (15). Interestingly, the antidote used in acetaminophen overdose, *N*-acetylcysteine, has been used successfully to treat pennyroyal toxicity.

2.2.8 Mistletoe

European mistletoe, *Viscum album*, is a hemi-parasitic evergreen found worldwide that grows on the upper branches of many trees and shrubs. Although the evergreen leaves of mistletoe are capable of photosynthesis, this parasite depends on its host for water and mineral nutrients. According to a popular Christmas custom (probably of Scandinavian origin), when two people meet under hanging mistletoe they are obliged to kiss.

European mistletoe is a poisonous plant and may cause gastrointestinal problems including stomach pain, and diarrhea if consumed. Despite its toxicity, the leaves and twigs of the plant have been used traditionally as a digestive aid, a heart tonic and a sedative. More recently, mistletoe extracts have gained popularity in the treatment of various cancers. The toxicity of the plant is related to compounds synthesized as part of its life cycle and to toxins it accumulates from its host. Endogenous mistletoe toxins are predominantly viscotoxins, lectins and a number of alkaline proteins. The concentrations of these compounds vary considerably between plants, which partially explains why some can be ingested without harm but others lead to severe toxicity. It should, however, be remembered that chemicals derived from the host may be involved in any observed toxicity – alkaloids, cardenolides, tannins, and terpenes of exogenous origin from the host have all been identified in mistletoe leaves and berries. Fortunately, there are only a few reported cases of mistletoe hepatotoxicity (16).

2.2.9 Miscellaneous other hepatotoxic herbs

There are many other herbs that may cause liver damage. Skullcap is a native North American plant that grows in rich woods, bluffs and along roadsides in wet ditches. The flowers bloom from May to August. Skullcap is used as a sedative and calming agent, and is recommended by herbalists to treat nervous tension, epilepsy and hysteria. The active components of skullcap are thought to be scutellarin, catalpol and various tannins. It is known by herbalists that skullcap may cause abortion if given to pregnant women. Another complication of skullcap supplement use may be liver toxicity. Several cases of hepatotoxicity involving skullcap in combination with another herbal product have been reported. One involved skullcap in combination with valerian and another involved skullcap in combination with gingko biloba. Gotu kola (Sanskrit: Mandukaparni, *Centella asiatica*) has long been used in Indian Ayurvedic medicine for treatment of hypertension and wound healing. This preparation contains pentacyclic triterpenic saponosides, asiaticoside and madecassoside, which may cause liver damage. One study included three case reports for women aged 61, 52 and 49 years who presented with high ALT (1193, 1694 and 324 U/L), ALP (503, 472 and 484 U/L) and bilirubin (4.23 mg/dL or 72.3 µmol/L, 19.89 mg/dL or 340.1 µmol/L, and 3.9 mg/dL or 66.7 µmol/L, respectively) (reference intervals see ▶Tab. 2.2). All patients improved after discontinuation of *C. asiatica* (17).

2.3 Conclusions

This chapter focuses on herbs and herbal products that have been associated with hepatotoxicity. The most commonly encountered herbal toxicity arises from consumption of kava products. Despite its well-documented toxicity in the medical literature, kava is still promoted by herbalists as a sleeping aid and calming agent and as an alternative to drug therapy. In early 2002, the FDA and Health Canada issued advisories that kava consumption should be discontinued until further information regarding its safety and potential for liver damage were determined. Mills and colleagues surveyed health food stores in Toronto after this advisory was issued and found that 22 out of 34 stores surveyed still recommended kava products for anxiety and only nine stores mentioned safety concerns related to kava. The authors concluded that federal advisories did not influence sales of kava (18). The risks in using kava, germander, comfrey, chaparral, various Chinese weight loss products and mistletoe have been established based on mechanistic studies and sufficient case reports, but the risk of hepatotoxicity following the use of skullcap and valerian is less well documented and the safety of these products is unclear at present.

Because abnormal function tests may be an early indication of liver damage due to certain herbal supplements, laboratory scientists, clinical chemists and pathologists are the first healthcare professionals who can intervene to advise clinicians to investigate what herbal supplements a patient is taking. Abnormal liver function tests in the absence of negative serology for hepatitis, cytomegaly and Epstein Barr virus are a strong indication of drug- or herb-induced liver damage. Clinicians are aware of drug-induced liver damage, but many patients do not disclose their use of herbal remedies, so physicians may be confused when an apparently healthy individual with no obvious cause of liver disease presents with abnormal liver function tests. Laboratorians and clinicians must work as a team to identify such herb-induced liver toxicity because early intervention and discontinuation of the toxic herb resolves liver toxicity with supportive therapy in most cases.

References

1. Kawaguchi, T., Harada, M., Arimatsu, H., Nagata, S., Koga, Y., Kuwahara, R. et al. (2004) Severe hepatotoxicity associated with a N-nitrosofenfluramine containing weight loss product. J. Gastroenterol. Hepatol. 19:349–350.
2. Jamieson, D.D., Duffield, P.H., Cheng, D., Duffield, A.M. (1989) Composition of central nervous system activity of the aqueous and lipid extract of kava (Piper methysticum). Arch. Int. Pharmacodyn. 301:66–80.
3. Whittaker, P., Clark, J.J., San, R.H., Betz, J.M., Seifried, H.E., de Jager, L.S. et al. (2008) Evaluation of commercial kava extracts and kavalactone standards for mutagenicity and toxicity using the mammalian cell gene mutation assay in L5178Y mouse lymphoma cells. Food Chem. Toxicol. 46:168–174.

4. Li, X.Z., Razman, I. (2010) Role of ethanol in kava hepatotoxicity. Phytother. Res. 24:475–480.
5. Escher, M., Desmeules, J. (2001) Hepatitis associated with kava, a herbal remedy. Br. Med. J. 322:139.
6. Kraft, M., Spahn, T.W., Menzel, J., Senninger, N., Dietl, K.H., Herbst, H. et al. (2001) Fulminant liver failure after administration of the herbal antidepressant kava-kava. Dtsch. Med. Wochenschr. 126:970–972 (in German).
7. Whitton, P.A., Lau, A., Salisbury, A., Whitehouse, J., Evans, C.S. (2003) Kava lactones and kava-kava controversy. Phytochemistry 64:673–679.
8. Teschke, R., Genthner, A., Wolf, A. (2009) Kava hepatotoxicity: comparison of aqueous, ethanolic, acetonic kava extracts and kava herbs mixtures. J. Ethnopharmacol. 123:378–384.
9. Alderman, S., Kailas, S., Goldfarb, S., Singaram, C, Malone, D.G. (1994) Cholestatic hepatitis after ingestion of chaparral leaves: confirmation by endoscopic retrograde cholangiopancreatography and liver biopsy. J. Clin. Gastroenterol. 19:242–247.
10. Sheikh, N.M., Philen, R.M., Love, L.A. (1997) Chaparral associated hepatotoxicity. Arch. Intern. Med. 157:913–919.
11. Yeong, M.L., Swinburn, B., Kennedy, M., Nicholson, G. (1990) Hepatic veno-occlusive disease associated with comfrey ingestion. J. Gastroenterol. Hepatol. 5:211–214.
12. Castot, A., Larrey, D. (1992) Hepatitis observed during a treatment with a drug or tea containing wild germander. Evaluation of 26 cases reported to the Regional Centers for Pharmacovigilance. Clin. Biol. 16:916–922 (in French).
13. Laliberte, L., Villeneuve, J.P. (1996) Hepatitis after use of germander, a herbal remedy. Can. Med. Assoc. J. 154:1689–1692.
14. Favreau, J.T., Ryu, M.L., Braunstein, G., Orshansky, G., Park, S.S., Coody, G.L. et al. (2002) Severe hepatotoxicity associated with use of the dietary supplement LipoKinetix. Ann. Intern. Med. 136:590–595.
15. Anderson, I.B., Mullen, W.H., Meeker, J.E., Khojasteh-Bakht, S.C., Oishi, S., Nelson, S.D. et al. (1996) Pennyroyal toxicity: measurement of toxic metabolites levels in two cases and review of literature. Ann. Intern. Med. 124:726–734.
16. Harvey, J., Colin-Jones, D.G. (1981) Mistletoe hepatitis. Br. Med. J. (Clin. Res. Ed.) 282:186–187.
17. Jorge, O.A., Jorge, A.D. (2005) Hepatotoxicity associated with the ingestion of Centella asiatica. Rev. Esp. Enferm. Dig. 97:115–124.
18. Mills, E., Singh, R., Ross, C., Ernst, E., Wilson, K. (2004) Impact of federal safety advisories on health food store advice. J. Gen. Med. 19:269–272.

3 Kelp and thyroid panel tests

3.1 Introduction

Kelp (seaweed) is a brown algae that grows under water along colder coastlines. It requires nutrient-rich shallow ocean water at or below 20°C (68°F). Kelps are varieties of brown algae of the order Laminariales and Fucales, with some species growing over 61 m (200 ft) long. Kelp is harvested, dried, ground and then used as food (primarily in Eastern Asia), as a fertilizer, and for sodium and potassium salts in industrial processes. Kelps are also a source of thickening agents and colloid stabilizers used in many commercial products. Kelp is a rich source of vitamins, including thiamine, riboflavin, pantothenic acid, ascorbic acid, niacin, folate, vitamin A and vitamin E. Various minerals such as calcium, magnesium, potassium, sodium, zinc, copper, manganese and selenium are also found in kelp. The United States Department of Agriculture (USDA) website does not list iodine among the nutrients in kelp. The nutritional value of kelp depends on the species, season, water temperature, and geographic area.

During the 19th century, kelp was used to obtain sodium bicarbonate by burning and was one of the major commercial sources of soda ash as well as thickening product to prepare jelly and salad dressing. Kelp ash was also used in soap and glass production.

3.2 Kelp and iodine content

Asian populations consume seaweed as food in various forms: raw in salads and as a vegetable, pickled with sauce or with vinegar, as a relish and in sweetened jellies, and as a vegetable soup. The average daily intake of iodine from seaweed ingestion by the Japanese population is 1.2 mg/day (1). The recommended daily allowance (RDA) for iodine is 150 µg for adults. Therefore, consumption of excess amounts of kelp may cause iodine excess and abnormal thyroid function.

There are many species of seaweed in Asia and around the world. The iodine content of kelp varies by species and there are seasonal variations as well. Some 99% of the total iodine in *Laminaria japonica* is soluble, whereas in other algae the soluble iodine content ranges from 16 to 41%. Therefore, iodine bioavailability from kelp is species-dependent and may vary from an average of 12.9 to 3040 µg/g of kelp, depending on the species (2). In a study of 35 species of Chinese marine algae, iodine content ranged from 12.9 to 5939 µg/g the highest concentrations were observed in March and gradually decreased to the lowest concentration, observed between May and December (3). Martinelango et al. analyzed 11 different seaweed species collected from the coast of Northeastern Maine and observed that the iodine content ranged from

16 to only 3134 mg/kg. *Laminaria* species had the highest iodine content and *Laminaria digitata*, the most commonly used seaweed for preparing kelp tablets, contained the highest of all (3134±15 mg/kg) on a dry weight basis (4). Teas et al. reported that the iodine content of seaweeds (including kelp) varied from 16 to 8165 mg/kg; the highest amount was found in a sample of kelp granules prepared from *L. digitata* harvested off the coast of Iceland. The kelp granules were made of dried and pulverized seaweed. The iodine content of seaweed varies with the age and condition of the plant, with iodine loss thought to occur when the plant is no longer growing. The iodine content of common edible seaweeds is water-soluble (99.2%) whereas the edible seaweed sargassum had only 40% water-soluble iodine (5).

3.3 Kelp as a herbal supplement

Although Asian populations, especially the Japanese, consume kelp both as a food and a traditional Oriental medicine, Western populations seeking the health benefits of traditional Chinese medicine prefer to consume kelp in the form of herbal supplements. Kelp, also known as seaweed, sea-tang, bladderwrack and sea wrack, has been used in traditional Oriental medicine for a long time to treat conditions arising from iodine deficiency (goiter) and other symptoms (►Tab. 3.1). Kelp supplements exist in many forms, including liquids, tablets, powders, capsules and gel caps. The supplements may have kelp as the main ingredient or a minor ingredient. Kelp-containing herbal supplements rarely list iodine as a component and provide an estimate of iodine content in each capsule. Because the iodine content of kelp depends on the species, the time of harvesting and the source, different kelp products produced by different manufacturers may vary widely in iodine content. In one study, the authors examined 96 prescriptions and 127 over-the-counter (OTC) prenatal multivitamins marketed in the USA for iodine content. Of these products, 27 prescriptions and 87 OTC products contained iodine. According to the labels, 89% of the products contained ≥150 µg of iodine per serving. Of these products, 42 had kelp as the source of the iodine, 67 had potassium iodide and five

Tab. 3.1: Indications for the use of kelp in traditional medicine

Goiter
Underactive thyroid
Blood purifier
Cleansing agent
Bone and joint pain
Antioxidant
Digestive problems
Anti-inflammatory agent
Rheumatoid arthritis

had another ingredient as the source. Sixty of the products were randomly selected and measured for iodine content and then compared to the label content claim. While those that contained potassium iodide provided ~75% of what was stated on their labels, the multivitamins that contained kelp had large variations in actual iodine content (6). An Internet search of dietary supplements containing kelp revealed 423 products containing kelp, 397 of which were dietary supplements originating from Asia, Europe and North America.

3.4 Case report

A 55-year-old man with a large mass on the right side of his neck was diagnosed with papillary thyroid carcinoma. The patient had a total thyroidectomy and pathological examination revealed a multifocal papillary thyroid carcinoma. Treatment was with radioiodine. For this procedure to be effective, plasma iodine concentrations must be reduced as nonradioactive iodine competes with radioactive iodine for uptake in thyroid tissue. A random urinary iodine concentration was obtained and was elevated at 394 µg/L (reference interval 42–350 µg/L). The patient's history revealed that he had ingested a large number of supplements; however the labels did not indicate an obvious source of iodine. A 24-h urine collection was then obtained, and the iodine content was again found to be elevated. The patient was given furosemide 20 mg daily for 10 days and placed on a low-iodine diet, but his 24-h urinary iodine excretion remained elevated. The supplements the patient was taking were again evaluated and it was found that the selenium supplement contained kelp which was listed as an inactive ingredient in the supplement. Kelp was the source of the excess iodine. After 8 weeks on a regular diet without supplements, the patient's 24-h urinary iodine excretion decreased to 192 µg. After consumption of a low-iodine diet for 4 weeks, 24-h urinary iodine excretion further decreased to 36 µg. Ablation with radioiodine was then performed successfully (7).

3.5 Role of iodine in thyroid function

The trace element iodine is essential for physiological functions because it is needed for synthesis of thyroid hormones; triiodothyronine (T_3) and thyroxine (T_4) . The thyroid gland has the capacity to trap iodine from the circulation and incorporate it in thyroid hormones, which, when released from the gland, can bind to thyroid receptors present in many target organs and tissues, triggering many essential physiological functions including growth, development, metabolism and even reproductive functions. Target tissues contain enzymes known as deiodinases that are capable of converting T_4 hormone, the most abundant thyroid hormone present in the blood, to T_3, which is the biologically active form of thyroid hormone. Thyroid hormone synthesis by the thyroid gland and levels of thyroid hormones in the blood are controlled by a complex feedback mechanism. The pituitary gland in the brain secretes thyroid-stimulating hormone (TSH) in response to thyroid-releasing hormone (TRH) secreted by the hypothalamus. TSH regulates trapping of iodine from the blood, as well as synthesis of thyroid hormones. When there are adequate levels of thyroid hormones in the blood, a feedback mechanism results in decreases in concentration of both TRH and TSH.

In the USA, the RDA for iodine is 150 μg/day for adults, but is higher for pregnant women (220 μg/day) and lactating mothers (290 μg/day). On average, Americans usually ingest 200 μg of iodine per day (8).

In healthy humans, the thyroid gland has an intrinsic autoregulatory mechanism to adapt to excess iodine present in the circulation. Conversion of elemental iodine to iodide (oxidation of iodine) is the first step in the synthesis of thyroid hormones and this process is inhibited if excess iodine is present in the circulation. This acute inhibitory effect of excess iodine in thyroid hormone production is known as the acute Wolff-Chaikoff effect and is only transient, lasting for approximately 48 h, after which normal production of thyroid hormone is restored. In euthyroid subjects, administration of up to 150 mg of iodine daily causes modest decreases in serum T_4 and T_3 concentrations, with compensatory increases in basal and TRH-stimulated serum TSH concentrations, but the values are usually within reference intervals. No change in thyroid hormone concentrations was observed on daily intake of up to 500 μg. Unfortunately, in the event of excess dietary iodine intake, some euthyroid patients with underlying thyroid disease may develop either hypothyroidism or hyperthyroidism (9).

Although normal individuals can ingest excess iodine without much harm, patients with an underlying thyroid disease may fail to adapt to excess dietary intake. This difference in physiological response to excess iodine is related to differences in iodide-induced inhibition of thyroid hormone synthesis. Subjects with elevated sensitivity to iodine, such as patients previously treated for Graves disease with radioactive iodine, subtotal thyroidectomy and antithyroid drugs and with those chronic autoimmune thyroiditis have an increased risk of developing hypothyroidism in response to excess dietary iodine because of a failure to escape from the Wolff-Chaikoff effect. By contrast, for subjects with lowered sensitivity, such as patients with goiter, goiter containing autonomous nodules and iodine deficiency, excess dietary iodine may cause hyperthyroidism. Iodine-induced hyperthyroidism has been reported in patients with iodine-deficiency goiter, patients with multinodular goiters who live in environments with low-iodine dietary sources, and patients with no evidence of underlying thyroid disease (10).

The etiology of iodine-induced hyperthyroidism is unclear, but is most commonly observed when iodine supplementation is given to iodine-deficient populations. Iodine-induced hyperthyroidism is far more common in areas of marginal iodine intake, such as continental Western Europe. It was also found in the US Midwest, an area with low dietary iodine intake until salt was supplemented with iodine in the 1920s. In the USA, where the population is aging and there is an associated increase in the frequency of nodular goiter, the incidence of iodine-induced hyperthyroidism is expected to increase, especially since iodine-containing substances, such as the iodine-rich contrast agents used in CT scans, are more frequently used in the elderly. Thyrotoxicosis is also observed in patients receiving the iodine-containing antiarrhythmic drug amiodarone due to a combination of excess iodine and amiodarone-induced inflammatory changes in the thyroid (9). Some euthyroid patients develop iodine-induced hypothyroidism with no apparent underlying thyroid disorder. These include children with cystic fibrosis, patients with chronic lung disease and the elderly. In the past, children with cystic fibrosis were often treated with iodide expectorants for their chronic pulmonary infections. Some of these iodide-treated children developed goiter and approximately 15% developed mild hypothyroidism. In a prospective study, goiter developed in approximately half of a cohort of cystic fibrosis children given large amounts of iodine for several weeks. None of these children had recognized underlying thyroid dysfunction. Patients with chronic lung disease and without evidence of underlying thyroid disease treated with expectorant iodine-containing medications showed an increased incidence of hypothyroidism. Iodinated glycerol had been advocated as an effective mucolytic expectorant in the treatment of lung disease, nasal congestion, and sinusitis. The use of iodinated glycerol in an elderly nursing home population induced hypothyroidism in a high percentage of patients and thyroid function returned to normal after the medication was withdrawn. It is possible that some of these elderly patients had mild subclinical hypothyroidism prior to administration of the iodinated glycerol, especially since mild elevations of serum TSH and thyroid antibodies are present in approximately 7% and 25%, respectively, of women in the USA over age 65 years (9).

Iodide readily crosses the placenta from mother to fetus. Administration of pharmacologic quantities of iodine-containing medications to pregnant and nursing women, either orally or locally (douching and during delivery), especially in regions of marginally low iodine intake, may induce transient hypothyroidism in the newborn. Topical application of the antiseptic povidone-iodine to the skin of newborns, especially low-weight infants, induces TSH elevations in approximately 25% of cases. Iodine contamination is the most common cause of transient neonatal hypothyroidism in continental Western Europe. Transient hypothyroidism has been reported in neonates born to mothers in Japan who had excess iodine intake due to seaweed ingestion (11).

3.6 Kelp and abnormal thyroid function

Mussig et al. reported a case involving a 39-year-old woman who had an enlarged thyroid gland. The patient's medical history and family history were unremarkable. Physical examination demonstrated a palpably enlarged thyroid and normal thyroid function tests, including TSH, free T_4 and free T_3, were congruent with the lack of clinical signs of hypothyroidism or hyperthyroidism. Antithyroid peroxidase (TPO), antithyroglobulin, and anti-TSH receptor antibody tests were also negative. Ultrasonography showed a multinodular goiter. The patient refused surgery and was advised to avoid excess intake of dietary iodine. Two months after her initial visit, the patient was doing well, but 4 months after her initial visit she presented with typical signs of hyperthyroidism, including tachycardia, palpitations, tremor, nervousness, insomnia, fatigue, increased sweating, diarrhea, secondary amenorrhea and weight loss. Laboratory tests showed increased levels of thyroid hormones (free T_4 3.2 ng/dL, free T_3 781 pg/dL) and a severely decreased level of TSH (<0.01 mU/L) (reference intervals: free T_4 0.8–2.4 ng/mL, free T_3 230–660 pg/dL, TSH 0.3–4.2 mU/L). An antithyroid antibody test was negative. The patient was not exposed to any iodine-containing drugs such as amiodarone or iodine-containing radiographic contrast agent. However, on further questioning, the patient revealed that she had taken a herbal tea for the previous 4 weeks, as prescribed by a Chinese herbalist to treat her enlarged thyroid gland. The tea contained large amounts of kelp, Kombu and Sargassum weed. It was estimated that the patient was ingesting between 580 and 990 μg of iodine daily, which was significantly greater than the RDA of 150 μg and led to iodine-induced hyperthyroidism. The patient was advised to discontinue the use of kelp tea and was treated with antithyroid medication thiamazole (40 mg daily) and propranolol (40 mg daily). A follow-up visit after 7 months revealed normal levels of thyroid hormones (free T_4 1.1 ng/dL, free T_3 280 pg/dL) and a slightly decreased TSH level (0.14 mU/L). Her thiamazole therapy was reduced to 20 mg per day (12).

De Smet et al. also reported a case of kelp-induced hyperthyroidism in a 50-year-old healthy woman who started taking six iodine-containing kelp tablets (200 mg of kelp each) daily to lose weight. She took no other medication. Within 2 months of taking the kelp supplement, she developed hyperthyroidism, which resolved after discontinuation of the supplement (13).

As mentioned earlier, excess iodine intake may cause hyperthyroidism or hypothyroidism. Clark et al. conducted a double-blind clinical trial in 36 euthyroid subjects who were randomly assigned to receive placebo, low-dose kelp (two tablets) or high-dose kelp (4 tablets) for 4 weeks. The authors observed that TSH levels did not change during the 4-week study in the placebo group, but slightly increased in the both low-dose and high-dose kelp groups. Total T_3 levels were significantly decreased in subjects receiving the high kelp dose, but did not change in the placebo or the low-dose kelp group. Urinary iodine excretion was increased in both the low-dose and high-dose kelp groups, as expected. The authors concluded that even short-term kelp supplement use may increase TSH levels (14). Konno et al. studied the association between dietary iodine intake and the prevalence of subclinical hypothyroidism in five coastal areas of Japan that produce iodine-rich seaweed (kelp). The relative frequency of high urinary iodine excretion in people living in these areas varied from 3.7 to 30.3%. The authors observed that hypothyroidism was more prevalent and marked in subjects consuming excess iodine, and excessive iodine intake should be considered as an etiological factor in hypothyroidism, in addition to chronic thyroiditis, in these areas (15).

3.7 Kelp and arsenic

Amster et al. reported a case involving a 54-year-old woman who was referred for evaluation of a 2-year history of worsening alopecia and memory loss. She also reported a skin rash, nausea and vomiting, which disabled her to an extent where she could no longer work full time. She admitted taking a daily kelp supplement. Her urine showed a very high level of arsenic (83.6 μg/g of creatinine) and analysis of the kelp supplement showed the presence of arsenic (8.5 mg/kg or 8.5 ppm). After discontinuation of the kelp supplement, her clinical condition improved significantly. The authors also analyzed commercially available kelp supplements and found arsenic at a level higher than approved by the US FDA (0.5–2 ppm is the level of arsenic acceptable in food supplements) in eight of the nine specimens analyzed (16).

3.8 Dietary iodine intake

Chronic iodine deficiency may lead to health problems in both adults and children, including thyroid gland dysfunction such as goiter, neurological dysfunction, and other health-related problems. Iodine deficiency in pregnant women may cause severe neurocognitive deficits in the newborn. However, iodine deficiency is rare in the USA and other developed countries owing to enrichment of table salt and cattle feeds with iodine. Humans mainly obtain iodine from their diet and the amount of iodine in food and water depends on the iodine content of the soil. However, iodine deficiency may be encountered in developing countries. Some vegetables contain iodine, but only if grown in iodine-rich soil. However, seawater is rich in iodine and iodine is abundant not only in seaweeds, but also in seafoods.

In Japan, where 21 species of seaweed are routinely included in the diet, and Korea, where more than 40 kinds of seaweed are used as food, excess iodine intake is common. In Hawaii and other Polynesian islands, more than 29 kinds of seaweeds are used as food, traditional medicines, and part of religious celebrations. Average Japanese seaweed consumption is 4–7 g/day and the estimated iodine intake in Japan varies from 200 to 20,000 µg/day (5). The iodine content of various foods is listed in ▶Tab. 3.2.

3.9 Conclusions

Kelp is an excellent source of fiber and nutrients, but there are consequences of kelp ingestion that patients and physicians should be made aware. Excess iodine ingestion due to use of a kelp supplement may cause hyper- or hypothyroidism, depending on the patient's sensitivity to excess dietary iodine. Iodine-induced thyroid disease is usually easily treated by removing the iodine source; radioiodine ablation therapy can be also be effective when the excess iodine source is identified and removed. Because patients often fail to disclose their use of herbal supplements, it is important to question patients regarding their use of supplements, especially in suspected cases of thyroid disorder due to excess iodine intake.

Tab. 3.2: Iodine content in various foods and kelp[a]

Food	Estimated iodine content for one serving (µg)
Iodized salt	77
Shrimp	35
Cod	99
Canned tuna	17
Cow's milk (1 cup, 8 oz)	56
Egg	12
Navy beans	32
Baked potato	34
Seaweed (variable)	>4500

[a]Source: Linus Pauling Institute, Oregon State University, http://www.lpi.oregonstate.edu/infocenter/minerals/iodine/, accessed May 30, 2010.

References

1. Nagataki, S. (2008) The average of dietary iodine intake due to the ingestion of seaweeds is 1.2 mg/day in Japan. Thyroid 18:667–669.
2. Hou, X., Chai, C., Qian, Q., Yan, X., Fan, X. (1997) Determination of chemical species of iodine in some seaweeds (I). Sci. Total Environ. 204:215–221.
3. Hou, X., Yan, X. (1998) Study on the concentration and seasonal variation of inorganic elements in 35 species of marine algae. Sci. Total Environ. 222:141–156.
4. Martinelango, P.K., Tian, K., Dasgupta, P.K. (2006) Perchlorate in seawater: bio-concentration of iodine and perchlorate in various seaweed species. Anal. Chim. Acta 567:100–107.
5. Teas, J., Pino, S., Critchley, A., Braveman, L.E. (2004) Variability of iodine content in common commercially available edible seaweed. Thyroid 14:836–841.
6. Leung, A.M., Pearce, E.N., Braverman, L.E. (2009) Iodine content of prenatal multivitamins in the United States. N. Engl. J. Med. 360:939–940.
7. Arum, S.M., He, X., Braverman, L.E. (2009) Excess iodine from an unexpected source. N. Engl. J. Med. 360:425–426.
8. Meletis, C.D., Zabriskie, N. (2007) Iodine, a critically overlooked nutrient. Altern. Compl. Ther. 135:132–136.
9. Braverman, L.E. (1994) Iodine and the thyroid: 33 years of study. Thyroid 4:351–356.
10. Roti, E., degli Uberti, E. (2001) Iodine excess and hyperthyroidism. Thyroid 11:493–499.
11. Nishiyama, S., Mikeda, T., Okada, T., Nakamura, K., Kotani, T., Hishinuma, A. (2004) Transient hypothyroidism or persistent hyperthyrotropinemia in neonates born to mothers with excessive iodine intake. Thyroid 14:1077–1083.
12. Mussig, K., Thamer, C., Bares, R., Lipp, H.P., Haring, H.U., Gallwitz, B. (2006) Iodine-induced thyrotoxicosis after ingestion of kelp-containing tea. J. Gen. Med. 21:C11–14.
13. de Smet, P.A., Stricker, B.H., Wilderink, F., Wiersinga, W.M. (1990) Hyperthyroidism during treatment with kelp tablets. Ned. Tijdschr. Geneeskd. 134:1058–1059 (in Dutch).
14. Clark, C.D., Bassett, B., Burge, M.R. (2003) Effect of kelp supplementation on thyroid function in euthyroid subjects. Endocr. Prac. 9:363–369.
15. Konno, N., Makita, H., Yuri, K., Iizuka, N., Kawasaki, K. (1994) Association between dietary intake and prevalence of subclinical hypothyroidism in the coastal regions of Japan. J. Clin. Endocrinol. Metab. 78:393–397.
16. Amster, E., Tiwary, A., Schenker, M.B. (2007) Case report: potential arsenic toxicosis secondary to herbal kelp supplement. Environ. Health Perspect. 115:606–608.

4 Interferences in digoxin immunoassays by various herbal supplements

4.1 Introduction

Only a relatively small number of drugs (~50) with a narrow therapeutic index require routine therapeutic drug monitoring and approximately 20–25 drugs are routinely monitored in most clinical laboratories. The International Association for Therapeutic Drug Monitoring and Clinical Toxicology adopted the following definition: 'Therapeutic drug monitoring is defined as the measurement made in the laboratory of a parameter that, with appropriate interpretation, will directly influence prescribing procedures. Commonly, the measurement is in a biological matrix of a prescribed xenobiotic, but it may also be of an endogenous compound prescribed as a replacement therapy in an individual who is physiologically or pathologically deficient in that compound' (1). Digoxin is a cardioactive drug requiring routine therapeutic monitoring owing to its very narrow therapeutic range of 0.8–1.8 ng/mL (1.0–2.3 nmol/L), although one clinical trial indicated that a beneficial effect of digoxin was observed at serum concentrations of 0.5–0.9 ng/ml (0.64–1.15 nmol/L), whereas serum concentrations $\geq$1.2 ng/mL (1.54 nmol/L) seem to be harmful (2). Because digoxin is present only in nanogram quantities in blood, it is more prone to interference from components of herbal supplements compared to other monitored drugs, which are usually present in serum in µg/mL amounts. For example, if an interfering compound is present in a herbal supplement at 50 ng/mL and has only 2% cross-reactivity with a digoxin immunoassay, it could falsely elevate serum digoxin values by up to 1 ng/mL, a very significant interference. Conversely, if this interfering substance also has 2% cross-reactivity with a phenytoin immunoassay, it could falsely elevate serum phenytoin levels by up to 1 ng/mL, but such interference would be insignificant because the phenytoin therapeutic range is 10–20 µg/mL (40–79 µmol/L).

4.2 Herbal supplements that interfere with digoxin immunoassays

Interference of herbal supplements with digoxin immunoassays can be classified under two broad categories:

- Class I: significant cross-reactivity with digoxin immunoassays. Examples are Chan Su, Lu-Shen-Wan (LSW), oleander and uzara roots. Such interference may cause confusion in interpreting digoxin results for both polyclonal and monoclonal digoxin immunoassays and may lead to clinically significant problems.
- Class II: moderate cross-reactivity with digoxin immunoassays. Examples are Asian and Siberian ginseng (see color plate), hawthorn and the Indian Ayurvedic medicine Ashwagandha. Interferences are usually negligible in monoclonal digoxin immunoassays and are only apparent if a polyclonal digoxin immunoassay is used.

4.2.1 Chan Su, Lu-Shen-Wan (LSW) and digoxin measurement

The Chinese medicine Chan Su is prepared from dried white secretions from the auricular glands and skin glands of Chinese toads (*Bufo melanostictus* Schneider and *Bufo bufo gargarzinas* Gantor). Chan Su is also a major component of other traditional medicines such as Lu-Shen-Wan and Kyushin. Despite their toxicity, these traditional Oriental medicines are used for the treatment of tonsillitis, sore throats and palpitations. The cardiotonic effect of Chan Su is due to its major bufadienolides such as bufalin (▶Fig. 4.1) which is structurally similar to digoxin, explaining the cswross-reactivity observed in various digoxin immunoassays (both monoclonal and polyclonal). Bufalin and related compounds are also found in aphrodisiacs sold under names such as Love Stone and Black Stone. Between February 1993 and May 1995, the New York City Poison Control Center was informed of the onset of illness in five previously healthy men after they ingested a topical aphrodisiac. Four men died from taking the substance. For example, one 26-year-old man started vomiting several hours after ingesting the topical aphrodisiac and his serum digoxin level was measured as 2.8 ng/mL (3.6 mmol/L) even though he never took digoxin (3).

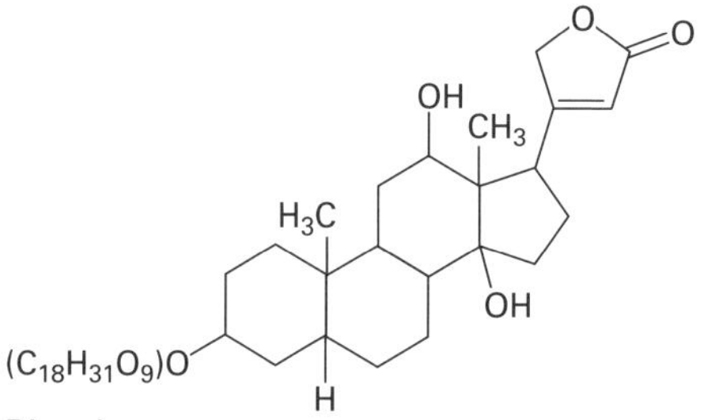

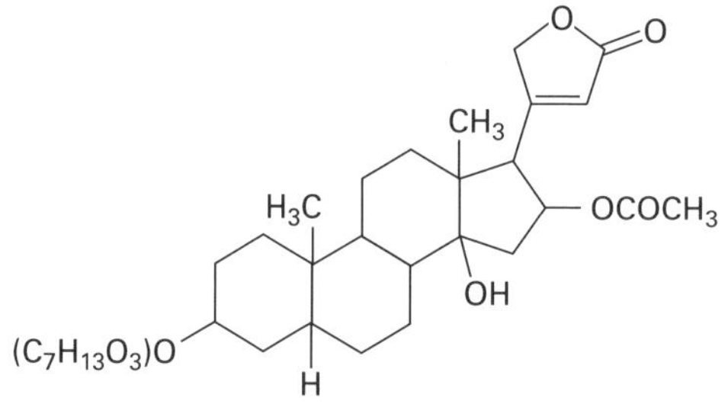

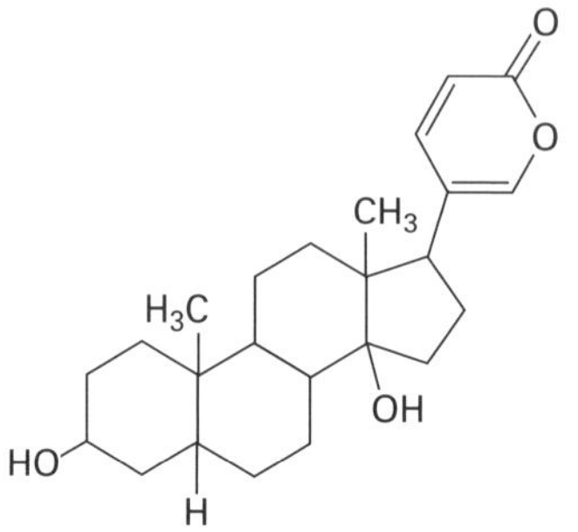

Fig. 4.1: Chemical structure of digoxin, oleandrin and bufalin.

The first report of interference by certain Oriental medicines in a serum digoxin immunoassay was published in 1992, when Panesar reported an apparent digoxin concentration of 1124 pmol/L (0.88 ng/mL) in a healthy volunteer who ingested LSW pills. Chan Su is the major component of LSW pills (4). An apparent serum digoxin concentration of 4.9 ng/mL (6.3 pmol/L) in the FPIA assay was reported for a woman who died from ingestion of Chinese herbal tea containing Chan Su (5). However, the fluorescence polarization immunoassay for digoxin (FPIA) for application on the TDX analyzer (Abbott Diagnostics, Abbott Park, IL) is no longer commercially available.

Although interference by Chan Su and related Oriental medicines prepared from toad venom in most digoxin immunoassays is positive (digoxin values are falsely increased), negative interference has also been reported for the Digoxin II assay (microparticle enzyme immunoassay, MEIA, manufactured by Abbott Diagnostics) (6). Negative analytical interference poses a greater clinical challenge than positive interference because a clinician may simply increase the dose of digoxin on the basis of a lower-than-expected digoxin serum concentration in a patient. However, for the later version of this assay (Digoxin III) on the AxSYM analyzer, Chan Su and related compounds showed positive rather than negative interference (7). Owing to the close structural similarity of digoxin and bufalin, monoclonal digoxin immunoassays such as the Tina Quant (Roche Diagnostics, Indianapolis, IN), Beckman Coulter Synchron digoxin assay (Beckman Coulter, Brea, CA) and EMIT digoxin assay are all affected by bufalin-containing Oriental medicines, but the magnitude of the interference is lower (25–50% lower, depending on the assay) than that observed for polyclonal digoxin immunoassays such as Digoxin III and FPIA.

The digoxin-like immunoreactive components of Chan Su and LSW significantly bind to serum proteins (>90%) and are mostly absent in protein-free ultrafiltrates. Therefore, free digoxin concentrations can be measured in protein-free ultrafiltrate when patients are taking these products to partly eliminate such interference in digoxin assays (6).

4.2.2 Oleander-containing products and digoxin immunoassays

The oleanders are evergreen ornamental shrubs belonging to the dogbane family that grow in Southern parts of the USA from Florida to California. Oleander plants also occur in various other parts of the world, including Australia, India, Sri Lanka and China. The most common oleander found in the USA is *Nerium oleander*, which has either pink or white flowers. Yellow oleander, also found in some parts of the USA but primarily in Asia, is a different species (*Thevetia peruviana*). Oleander plants are among the most toxic plants known and all the plant parts are toxic. The toxic effect of oleander can occur on exposure to a small amount of the plant. Boiling or drying of the plant does not inactivate the toxins. Human exposure to both pink and yellow oleander includes accidental exposure, ingestion by children, administration in food or drink, medicinal preparations from oleander (herbal products) and criminal poisoning. Deliberate ingestion of oleander seeds is also a popular method of suicide in Sri Lanka and other parts of the world. Despite its known toxicity, oleander is used in folk medicines, and death from drinking a herbal tea containing oleander has been reported (8).

Cardiac glycosides are found in both pink and yellow oleander. The major cardiac glycoside found in pink oleander is oleandrin (▶Fig. 4.1). Yellow oleander contains at least eight different cardiac glycosides, including thevetin A, thevetin B, thevetoxin, neriifolin, peruvoside and ruvoside. Rosagenin from the bark of the plants exhibits a strychnine-like effect.

Cardiac glycosides present in both pink and yellow oleander cross-react with digoxin immunoassays. Osterloh et al., using a digoxin radioimmunoassay that is no longer commercially available, observed a serum digoxin level of 5.8 ng/mL in a woman who died after ingesting oleander leaves. Based on the cross-reactivity of oleander extract in the method, the authors estimated that the woman ingested 4 g of oleander leaves (9). Haynes et al. reported a case in which a woman died after drinking herbal tea prepared from pink oleander (8). Eddleston et al. reported a mean apparent serum digoxin concentration of 1.2 ng/mL (1.5 nmol/L) in patients who were poisoned with yellow oleander but eventually discharged from hospital. Severe oleander toxicity resulted in a mean apparent serum digoxin concentration of 2.2 ng/mL (2.8 nmol/L) as measured by the FPIA digoxin assay (10).

Oleander poisoning (both pink and yellow) can be indirectly detected by FPIA on a TDx analyzer, but Abbott Diagnostics discontinued this assay in May 2009. In our experience, the FPIA assay showed the highest cross-reactivity with oleander, although the Beckman Coulter Synchron digoxin assay on a Synchron LX analyzer also showed some interference by oleander. However, the magnitude of the interference was approximately 65% less for the Beckman Coulter than for the FPIA assay (11). More recently, Abbott Diagnostics marketed the Digoxin III assay for measurement of digoxin concentrations on the AxSYM analyzer. This new assay has sensitivity similar to that of the FPIA assay in detecting oleander poisoning (12). We recently observed that the Siemens Dimension Vista digoxin assay has similar sensitivity to the FPIA assay in detecting oleander poisoning. When an aliquot of drug-free serum was spiked with 100 ng/mL of oleandrin, the apparent digoxin concentration was 1.2 ng/mL (1.5 nmol/L) according to the Dimension Vista digoxin assay and 1.1 ng/mL (1.4 nmol/L) according to the FPIA assay. By contrast, the Tina Quant monoclonal digoxin immunoassay showed no apparent digoxin at this oleandrin concentration. However, when an aliquot of drug-free serum was spiked with oleandrin to achieve a final concentration of 1 µg/mL, the apparent digoxin concentration was 2.2 ng/mL (2.8 nmol/L) using the Dimension Vista assay, 2.0 ng/mL (2.6 nmol/L) using the FPIA assay and 0.7 ng/mL (0.9 nmol/L) using the Tina Quant assay (Dasgupta et al, unpublished data).

Oleandrin strongly binds to serum protein and is absent in the protein-free ultrafiltrate. Therefore, monitoring of free digoxin in protein-free ultrafiltrate may eliminate some interference by oleander in serum digoxin assays provided oleandrin concentrations are low to moderate.

4.2.3 Uzara root and digoxin immunoassays

Thurmann et al. reported that glycosides from uzara root may interfere with serum digoxin immunoassays. The authors investigated digoxin and digitoxin concentrations after four healthy volunteers ingested 1.5 mL (~22 drops) of uzara. Maximum digoxin concentrations of 1.1–4.9 ng/mL (1.4–6.34 µmol/L) were observed 6 h after dosing (13). Although popular in Germany, uzara root is not readily available in the USA.

4.2.4 Various ginsengs and digoxin immunoassays

Ginsengs (see color plate) are slow-growing perennial plants found mostly in cooler regions of Eastern Asian, such as China, Korea and Eastern Siberia. Ginseng refers to 11 species within the *Panax* genus. Ginsengs have been used in traditional Chinese medicine for over 2500 years and are considered capable of curing many illnesses. Extracts from the roots of Asian ginseng (*Panax ginseng*) and American ginseng (*Panax quinquefolius*) are normally used in the USA. Siberian ginseng (*Eleutherococcus senticosus*) is not a true ginseng because it does not contain characteristic ginsenosides; instead it contains eleutherosides. Ginseng is promoted as a tonic for general wellbeing. The desired effects are increased energy and enhanced sexual function, but it is hard to conduct clinical trials to determine ginseng efficacy because of the subjective nature of feeling more energetic or calms. The term Indian ginseng is a misnomer because it is prepared from an entirely different plant, *Withania somnifera*. This herbal product is known as "Ashwagandha" in traditional Indian Ayurvedic medicine and has been used for over 3000 years as an aphrodisiac, liver tonic, anti-inflammatory agent and astringent. The major biochemical constituents of Ashwagandha are steroidal alkaloids and steroidal lactones in a class of compounds termed withanolides, which have some structural similarity to digoxin (►Fig. 4.1).

In 1996, McRae reported a case of a 74-year-old man with a steady serum digoxin level of 0.9–2.2 ng/mL (1.1–2.8 nmol/L) for 10 years, which increased on one occasion to 5.2 ng/mL (6.6 nmol/L) after taking Siberian ginseng. Although the level was toxic, the patient did not experience any signs or symptoms of digoxin toxicity. The patient stopped taking Siberian ginseng and his digoxin level returned to the therapeutic range (14). However, in our experience Siberian ginseng only shows very modest interference with the FPIA assay, which is no longer commercially available, and most digoxin assays we tested showed no effect at all. Therefore, it is possible that in this case the patient ingested some other herbal remedy mislabeled as Siberian ginseng. Mislabeling of Chinese herbs has previously been reported.

Asian ginseng, which is prepared from a different herb than Siberian ginseng, also showed modest interference with the FPIA assay, but not with other digoxin immunoassays based on more specific monoclonal antibodies. Similarly, Ashwagandha showed very modest interference with the FPIA assay, but not with other digoxin immunoassays such as the Tina Quant assay and the Beckman Coulter assay on the Synchron LX analyzer (15).

4.2.5 DanShen

DanShen is a Chinese medicine prepared from the root of the herb *Salvia miltiorrhiza*. This drug has been used in China for many years in the treatment of various cardiovascular diseases, including angina pectoris, and is readily available in the USA in Chinese herbal stores. More than 20 diterpene quinones known as tanshinones have been isolated from DanShen. These compounds have some structural similarity to digoxin, but interference was only observed for the FPIA assay, which is no longer commercially available (16). Interference by DanShen with monoclonal-antibody-based assays such as the Tina Quant, Advia (Siemens, Tarrytown, NY) and Beckman Coulter Synchron digoxin assays is non-significant.

4.3 Conclusions

At present, interferences in immunoassays for therapeutic drugs due to ingestion of herbal remedies have been observed only for digoxin assays. The magnitude of interference

Tab. 4.1: Herbal supplements that significantly interfere with digoxin immunoassays

Herbal product	Interference	Comments
Chan Su Lu-Shen-Wan (LSW)	Monoclonal and polyclonal	Falsely elevates serum digoxin values but falsely lowers digoxin values (negative interference) in the MEIA digoxin assay; monitoring of free digoxin may eliminate interference if bufalin concentration is moderate, as expected after taking moderate amounts of such Oriental medicines
Oleander	Monoclonal and polyclonal	Falsely elevates serum digoxin values but falsely lowers (negative interference) digoxin values in the MEIA Digoxin assay by Abbott Laboratories; monitoring of free digoxin may eliminate interference if oleandrin is present in modest amounts
Uzara root	Polyclonal	Interferes with digoxin assays, but the effect on monoclonal antibody-based immunoassays has not been studied, or the protein binding of digoxin-like immunoreactive components of uzara root

is high for Chinese medicines Chan Su and LSW, as well as for oleander-containing herbal products (▶Tab. 4.1) because bufalin (active component of Chan Su and LSW) and oleandrin (active component of oleander extract) are structurally very similar to digoxin (▶Fig. 4.1). These herbal products affect both polyclonal and monoclonal digoxin immunoassays, although the magnitude of interference is more significant for polyclonal than for monoclonal immunoassays. Various Indian Ayurvedic medicines, ginsengs and DanShen only show moderate interference in polyclonal-antibody-based digoxin immunoassays such as FPIA (no longer available) and MEIA assays (old Digoxin II and relatively new Digoxin III). More specific monoclonal-antibody-based digoxin assays are not affected by Ashwagandha, Asian ginseng, Siberian ginseng and DanShen. Therefore, if a clinical laboratory uses a monoclonal-antibody-based immunoassay for routine therapeutic drug monitoring of digoxin, the chance of falsely elevated digoxin values is minimized, but significant interference may still occur in a patient on digoxin therapy and taking Chan Su and related Oriental medicines or oleander-containing herbal supplements.

References

1. Watson, I., Potter, J., Yatscoff, R., Fraser, A., Himberg, J.J., Wenk, M. (1997) Editorial. Ther. Drug Monit. 19:125.
2. Adams, K.F., Patterson, J.H., Gattis, W.A., O'Connor, C.M., Lee, C.R., Schwartz, T.A. et al. (2005) Relationship of serum digoxin concentrations to mortality and morbidity in women in the Digitalis Investigation Group trial: a retrospective study. J. Am. Coll. Cardiol. 46:497–504.
3. Center for Disease Control and Prevention. (1995) Deaths associated with a purported aphrodisiac – New York City, February 1993–May 1995. Morbid. Mortal. Wkly. Rep. 44:853–855.

4. Panesar, N.S. (1992) Bufalin and unidentified substances in traditional Chinese medicine cross-react in commercial digoxin assay. Clin. Chem. 38:2155–2156.

5. Ko, R., Greenwald, M., Loscutoff, S., Au, A., Appel, B.R., Kreutzer, R.A. et al. (1996) Lethal ingestion of Chinese tea containing Chan SU. West. J. Med. 164:71–75.

6. Dasgupta, A., Biddle, D., Wells, A., Datta, P. (2000) Positive and negative interference of Chinese medicine Chan SU in serum digoxin measurement: elimination of interference using a monoclonal chemiluminescent digoxin assay or monitoring free digoxin concentrations. Am. J. Clin. Pathol. 114:174–179.

7. Reyes, M.A., Actor, J.K., Risin, S.A., Dasgupta, A. (2008) Effect of Chinese medicine Chan Su and Lu-Shen-Wan on serum digoxin measurement by Digoxin III, a new digoxin immunoassay. Ther. Drug Monit. 30:95–99.

8. Haynes, B.E., Bessen, H.A., Wightman, W.D. (1985) Oleander tea: herbal draught of death. Ann. Emerg. Med. 14:350–353.

9. Osterloh, J., Harold, S., Pond, S. (1982) Oleander interference in the digoxin radioimmunoassay in a fatal ingestion. J. Am. Med. Assoc. 247:1596–1597.

10. Eddleston, M., Ariaratnam, C.A., Sjostrom, L., Jayalath, S., Rajakanthan, K., Rajapakse, S. et al. (2000) Acute yellow oleander (Thevetia peruviana) poisoning: cardiac arrhythmias, electrolyte disturbances, and serum cardiac glycoside concentrations on presentation to hospital. Heart 83:310–306.

11. Dasgupta, A., Datta, P. (2004) Rapid detection of oleander poisoning by using digoxin immunoassays: comparison of five assays. Ther. Drug Monit. 26:658–663.

12. Dasgupta, A., Risin, S.A., Reyes, M., Actor, J.K. (2008) Rapid detection of oleander poisoning by Digoxin III, a new digoxin assay: impact on serum digoxin measurement. Am. J. Clin. Pathol. 129:548–553.

13. Thurmann, P.A., Neff, A., Fleisch, J. (2004) Interference of Uzara glycosides in assays of digitalis glycosides. Int. J. Clin. Pharmacol. Ther. 42:281–284.

14. McRae, S. (1996) Elevated serum digoxin levels in a patient taking digoxin and Siberian ginseng. Can. Med. Assoc. J. 155:293–295.

15. Dasgupta, A., Peterson, A., Wells, A., Actor, J.K. (2007) Effect of Indian Ayurvedic medicine Ashwagandha on measurement of digoxin and 11 commonly monitored drugs using immunoassay: study of protein binding and interaction with Digibind. Arch. Pathol. Lab. Med. 131:1298–1303.

16. Wahed, A., Dasgupta, A. (2001) Positive and negative in vitro interference of Chinese medicine Danshen in serum digoxin measurement: elimination of interference by monitoring free digoxin concentrations. Am. J. Clin. Pathol. 116:403–408.

5 Interaction of St. John's wort with various drugs

5.1 Introduction

Complementary and alternative medicines are becoming increasingly popular in the USA, Europe and other parts of the world. The general concept often portrayed in marketing material and the media that anything natural is safe is not true. Herbal remedies can be toxic and inappropriate use or overuse may even cause fatality. Death following the use of kava and related hepatotoxic herbs has been discussed in chapter 2. Sometimes a relatively safe herbal supplement such as St. John's wort (see color plate) can interact with Western drugs and cause treatment failure. Karhova et al. described the case of a 63-year-old woman who received a liver allograft for cryptogenic liver cirrhosis. The patient developed acute organ rejection 14 months after the transplant due to a sudden decrease in whole-blood cyclosporine A to 48 ng/mL, a sub-therapeutic level (Therapeutic concentration 75–300 ng/mL depending on transplant type. If multiple immunosuppressants are used, lower level may be targeted.). A study of cyclosporine A resorption showed increased metabolism of the drug and the dose was increased from 175 mg twice daily to 300 mg twice daily to achieve therapeutic levels. The rejection episode was controlled by a steroid bolus. The authors discovered that the woman had taken *Hypericum perforatum* (St. John's wort, 900-mg capsule twice a day) for 2 weeks prior to her rejection episode to counteract increasing depression. The patient was advised to discontinue St. John's wort and her cyclosporine A level increased. Eventually her cyclosporine dose was reduced to 150 mg twice daily (close to her original dosage) but her whole-blood cyclosporine A level was within the therapeutic range (125–150 ng/mL). This case illustrates how a relatively safe herbal supplement can cause a clinically adverse, potentially life-threatening situation due to drug interaction (1). Besides cyclosporine A, St. John's wort interacts with many other Western drugs and causes clinically significant effects. Therefore, patients taking medications to treat a chronic condition must avoid St. John's wort. Otherwise, St. John's wort is a relatively safe herbal supplement with efficacy in controlling mild depression (see chapter 1).

5.2 Mechanism of interaction between St. John's wort and Western drugs

Most commercially available St. John's wort preparations in the USA are dried alcoholic extracts or liquid extracts of the plant. Many chemicals have been isolated from St. John's wort, including hypericin, pseudohypericin, quercetin, isoquercitrin, rutin, amentoflavone, hyperforin, and other flavonoids and xanthenes. Hypericin, hyperforin and 1,3,5,7-tetrahydroxyxanthone are unique to St. John's wort.

The interaction of St. John's wort with a Western drug can be either a pharmacokinetic or a pharmacodynamic interaction, although pharmacokinetic interactions are more common. A pharmacokinetic interaction between a Western drug and St. John's wort involves the following mechanisms:

1. The primary mechanism of drug-herb interactions involves induction or inhibition of hepatic or intestinal metabolism of drugs by cytochrome P450 (CYP). CYP3A4 is the most abundant isoenzyme and is responsible for metabolism of more than 73 drugs and numerous endogenous compounds (2). The active components of St. John's wort induce CYP3A4, CYP2E1and CYP2C19. In particular, hyperforin is thought to be responsible for isoenzyme induction through activation of a nuclear steroid/pregnane and xenobiotic receptor (3). Induction of these liver enzymes accelerates the metabolism of many drugs following intake of St. John's wort, causing significant reductions in blood levels of these drugs and thus treatment failures. The chemical structure of hyperforin is shown in ▶Fig. 5.1.

2. Another mechanism of drug-herb interactions involves induction or inhibition of intestinal drug efflux pumps, including P-glycoprotein and multiple drug resistance proteins (MRP). Induction of efflux proteins may lead to significantly reduced concentrations of a drug in blood and tissue, causing treatment failure. St. John's wort also induces P-glycoprotein drug transporter and may reduce the efficacy of drugs for which hepatic metabolism is not the major clearance pathway. Hypericin may be the active ingredient in St. John's wort that modulates P-glycoprotein (4).

Self-medication with St. John's wort may cause treatment failures due to significant reductions in plasma drug concentrations because of increased clearance. Published reports indicate that St. John's wort significantly reduces steady-state plasma concentrations of many drugs, including alprazolam, amitriptyline, atorvastatin, chlorzoxazone, cyclosporine, debrisoquine, digoxin, erythromycin, fexofenadine, gliclazide, imatinib, indinavir, irinotecan, mephenytoin, methadone, midazolam, nifedipine, omeprazole, oral contraceptives, quazepam, simvastatin, tacrolimus, verapamil, voriconazole and warfarin. Case reports also suggest that St. John's wort interacts with bupropion, loperamide, nefazodone, nevirapine, paroxetine, phenprocoumon, prednisone, sertraline and venlafaxine (5). A majority of the interactions between St. John's wort and various Western drugs are pharmacokinetic in nature, although several clinically significant pharmacodynamic interactions have also been reported. Clinically significant pharmacokinetic interactions between St. John's wort and various Western drugs are listed in ▶Tab. 5.1. Drugs that interact pharmacodynamically with St. John's wort and drugs that unexpectedly show no interaction are listed in ▶Tab. 5.2.

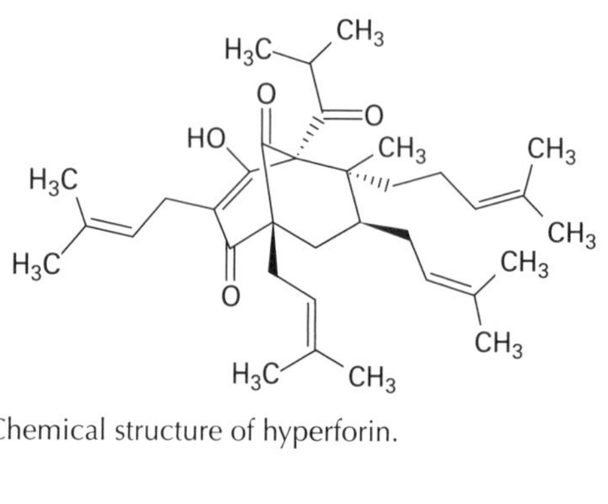

Fig. 5.1: Chemical structure of hyperforin.

Tab. 5.1: Clinically significant pharmacokinetic interactions between St. John's wort and various drugs

Class	Drug name
Immunosuppressants	Cyclosporine, tacrolimus
Anticoagulant	Warfarin
Antiretroviral agent	Indinavir, saquinavir, atazanavir, lamivudine, nevirapine
Anticancer agents	Imatinib, irinotecan
Cardiovascular drugs	Digoxin, verapamil, nifedipine
Benzodiazepines	Alprazolam, midazolam, quazepam
Antiepileptic	Mephenytoin, phenobarbital
Hypoglycemic agent	Gliclazide
Antimicrobial agents	Erythromycin, voriconazole
Anti-asthmatic agent	Theophylline
Proton pump inhibitor	Omeprazole
Statins	Simvastatin, atorvastatin
Oral contraceptives	Ethinylestradiol and related compounds
Antidepressant	Amitriptyline
Synthetic opioids	Methadone, oxycodone

Therapeutic efficacy of all these drugs is reduced due to interaction with St. John's wort.

Tab. 5.2: Drugs that interact pharmacodynamically with St. John's wort or unexpectedly show no interaction

Interaction	Drug
Pharmacodynamic interaction (causing serotonin syndrome)	Paroxetine
	Sertraline
	Venlafaxine
	Nefazodone
Potential pharmacodynamic interaction	Cocaine
No interaction	Carbamazepine
	Mycophenolic acid
	Pravastatin
	Tolbutamide

5.2.1 Hyperforin content and interaction extent

The hyperforin content determines the magnitude of interaction between a Western drug and St. John's wort. Hypericin and hyperforin are the most studied components of St. John's wort, and evidence suggests that hyperforin mediates most or all of the anti-depressive effects of the herb. Most commercial preparations of St. John's wort are standardized to 0.3% hypericin, which usually corresponds to approximately 3% hyperforin, but concentrations can vary in preparations sold to the public (6). In one report, the authors observed that hyperforin amounts in a commercially available St. John's wort preparation could vary 62-fold, ranging from 0.49 to 39.57 mg of hyperforin per dose. The magnitude of cytochrome P450 induction, especially for CYP3A4, is directly proportional to the hyperforin content (7). Madabushi et al. studied the correlation between hyperforin content and the magnitude of interactions with Western drugs and observed that St. John's wort preparations containing <1% hyperforin do not demonstrate clinically significant drug-herb interactions. Commercial St. John's wort preparations usually contain 3% hyperforin. Interestingly, although low-hyperforin St. John's wort may not cause clinically significant drug interactions, such preparations clearly demonstrated superiority over placebo and were equivalent to imipramine and fluoxetine in the treatment of mild to moderate depression (8). However, many commercial preparations of St. John's wort contain more than 1% hyperforin and will result in clinically significant interactions with various Western drugs. Interaction with certain drug classes and warfarin could have very serious clinical consequences.

5.3 Interaction between immunosuppressants and St. John's wort

Interaction between St. John's wort and various immunosuppressants is clinically very significant because the herb may reduce trough concentrations of cyclosporine A and tacrolimus by almost 50%, which could cause life-threatening organ rejection in an otherwise stable transplant recipient. Interestingly, there is no reported interaction between St. John's wort and mycophenolic acid, an immunosuppressant.

Barone et al. reported two cases in which renal transplant recipients started self-medication with St. John's wort. Both patients experienced sub-therapeutic concentrations of cyclosporine and one patient developed acute graft rejection due to low cyclosporine concentrations. Both patients were advised to discontinue the use of St. John's wort immediately and their cyclosporine A concentrations returned to therapeutic levels (9). Alschner and Klotz reported a case study of a 57-year-old kidney transplant recipient on a long-term regular doses of cyclosporine A (125–150 mg/day) and prednisolone (5 mg/day). Routine monitoring of cyclosporine A trough levels (100–130 ng/mL) over the previous 2 years revealed a sudden decrease to 70 ng/mL despite an increase in cyclosporine A to 250 mg/day. The patient admitted to taking a herbal tea mixture for depression that contained St. John's wort. Five days after discontinuing the herbal tea, his cyclosporine level increased to170 ng/mL on a dose of 250 mg/day. The cyclosporine A dose was then reduced to 175 mg/day and his trough level remained at approximately 130 ng/mL (10).

Mai et al. reported that the hyperforin content of St. John's wort determines the magnitude of interaction with cyclosporine. The authors ground up commercial St. John's wort tablets and half of the resulting powder was subjected to supercritical carbon dioxide extraction to remove hyperforin, the active component of St. John's wort responsible for interaction with cyclosporine A. Hard gelatin capsules were subsequently prepared with both preparations (low and high hyperforin) to contain 150 mg of St. John's wort extract per capsule. The original powder contained 7.0 mg of total hyperforin, 0.45 mg of total hypericin, and 16.16 mg of total flavonoids per capsule, whereas the low-hyperforin powder contained 0.1, 0.45 and 15.6 mg, respectively. Therefore, the original St. John's wort extract contained approximately 4.6% hyperforin, whereas the low-hyperforin preparation had 0.6% hyperforin. The authors performed a crossover study in 10 renal transplant patients randomized to two groups that received either the normal or the low-hyperforin St. John's wort extract for 14 days in addition to their regular regimen of cyclosporine A. After a 27-day washout phase, patients were crossed over to the other treatment group for 14 days. Blood cyclosporine A concentrations were measured by an immunoassay based on a specific monoclonal antibody against cyclosporine A. The authors observed significant reductions in cyclosporine A [0–12-h area under the curve (AUC)] in patients receiving the normal St. John's wort (52% reduction compared to baseline), requiring a 65% increase in daily cyclosporine A dose to maintain therapeutic levels. By contrast, no dose increase was necessary for patients receiving the low-hyperforin preparation because no significant reduction in cyclosporine A levels in blood compared to baseline was observed (11).

A significant reduction in AUC for tacrolimus was observed in ten stable renal transplant patients taking St. John's wort. The peak tacrolimus concentration also decreased from a mean of 29.0 ng/mL to 22.4 ng/mL (therapeutic concentration 5–15 ng/mL) following co-administration of St. John's wort (12). Bolley et al. reported on a 65-year-old patient who received a renal transplant in November 1998 and had a trough whole-blood level tacrolimus concentrations between 6 and 10 ng/mL. The patient started self-medication with St. John's wort in July 2000 (600 mg/day) because of depression and in August 2000 showed an unexpected low tacrolimus concentration of 1.6 ng/mL. When the patient stopped taking St. John's wort, tacrolimus levels returned to the previous range (13). Mai et al. studied the interaction of St. John's wort with tacrolimus and mycophenolic acid in 10 stable renal transplant patients. Co-administration of St. John's wort significantly reduced AUC and both peak and trough blood concentrations of tacrolimus. To achieve sufficient immunosuppression, tacrolimus doses were increased in all patients (median dose increased from 4.5 to 8.0 mg/day). Interestingly, the pharmacokinetic parameters of mycophenolic acid were not affected by co-administration of St. John's wort (14).

5.4 Interaction between antiretrovirals and St. John's wort

St. John's wort should not be combined with antiretroviral therapy, including highly active antiretroviral therapy (HAART). This warning follows a study by the National Institutes of Health (Department of Health and Human Services, United States Government) on dosing of blood levels of indinavir in HIV-negative test subjects (15). In addition to St. John's wort, patients suffering from AIDS and being treated with antiretroviral agents should not take echinacea (see color plate), garlic (see color plate), ginkgo (see color plate) or milk thistle because of interactions between these herbal remedies and antiretroviral drugs (16).

In one study, St. John's wort reduced AUC for the HIV-1 protease inhibitor indinavir by a mean of 57% and decreased the extrapolated trough by 81%. The subjects received 300 mg of St. John's wort three times a day for 14 days. The mean peak indinavir concentration decreased from 12.3 to 8.9 ng/mL (therapeutic concentration 80–120 ng/mL) in healthy volunteers taking both indinavir and St. John's wort. A more significant effect was observed for the C8 metabolite of indinavir, for which the mean concentration decreased from 0.494 ng/mL in the control group to 0.048 ng/mL in the St. John's wort group (17). Busti et al. reported that atazanavir therapy can also be affected by simultaneous use of St. John's wort (18). Co-administration of lopinavir/ritonavir and St. John's wort resulted in a substantial reduction in lopinavir plasma concentrations (19). Nevirapine, a non-nucleoside reverse transcriptase inhibitor, is metabolized by CYP3A4 and CYP2D6. Increased oral nevirapine clearance following St. John's wort administration has been reported in five HIV patients (20).

5.5 Interaction between warfarin and St. John's wort

Warfarin is an anticoagulant used in treating patients with venous thrombosis and pulmonary embolism. Warfarin is also used to prevent blood clots in patients with atrial fibrillation. Concurrent administration of St. John's wort and warfarin reduces the efficacy of the latter due to increased warfarin clearance. Therefore, patients on warfarin therapy must avoid St. John's wort. In addition, warfarin has significant interactions with many herbal supplements, including garlic (see color plate), ginkgo (see color plate), ginseng (see color plate), cranberry juice and vitamin E supplements. These important interactions are discussed in detail in chapter 6.

5.6 Interaction between anticancer drugs and St. John's wort

Considering the narrow therapeutic window of anticancer drugs, interaction with St. John's wort and any subsequent decrease in serum drug level is of particular clinical significance. Clearance of imatinib mesylate, an anticancer drug, increases on administration of St. John's wort and results in reduced clinical efficacy of the drug. Imatinib is used in the treatment of Philadelphia chromosome positive chronic myeloid leukemia and gastrointestinal stromal tumors. In one study involving ten healthy volunteers, administration of St. John's wort for 2 weeks significantly reduced maximum plasma imatinib concentrations by 29% and AUC by 32%. In addition, the half-life of the drug was reduced by 21% (21). St. John's wort also showed significant interaction with another anticancer drug, irinotecan. In one study involving five patients, ingestion of St. John's wort (900 mg/day) for 18 days resulted in an average 42% reduction in concentrations of SN-39, the active metabolite of irinotecan. This reduction also caused a decrease in myelosuppression indicative of treatment failure (22).

5.7 Interaction between cardiovascular drugs and St. John's wort

Interactions between several medications used to treat various cardiovascular symptoms and St. John's wort may have clinical significance. Interaction between St. John's Wort and digoxin is well documented in the literature. Johne et al. reported that 10 days of St. John's wort use resulted in a 33% decrease in peak and 26% decrease in trough serum digoxin concentrations. The mean peak digoxin concentration was 1.9 ng/mL in the placebo group and 1.4 ng/mL the group taking St. John's wort. $AUC_{0-24\,h}$ was 25% lower in the group consuming St. John's wort compared to the placebo group (23). Digoxin is a substrate for P-glycoprotein, which is induced by St. John's wort. Durr et al. confirmed that lower digoxin concentrations occur in healthy volunteers concomitantly taking St. John's Wort (24). In addition to digoxin, clinically significant interaction with St. John's wort has been reported for nifedipine and verapamil; both drugs are metabolized by CYP3A4 (5).

Procainamide, another cardioactive drug, is metabolized in the liver by N-acetyltransferase. Interaction between procainamide and St. John's wort was studied in Swiss Webster mice and it was observed that, unlike CYP3A4, components of St. John's wort have no inducing effect on N-acetyltransferase. When mice fed with St. John's wort for 2 weeks and were administered a single dose of procainamide 24 h after the last dose of St. John's wort, no difference in peak procainamide was observed between the St. John's wort group and the control group. In addition, the half-life of procainamide was not significantly different between the control and experimental groups. By contrast, when a group of mice was fed with a single dose of St. John's wort 1 h before administration of a standard procainamide dose, a significant increase in peak procainamide concentration was observed, but there was no change in procainamide half-life compared to the control group. When mice were fed with pure hypericin 1 h before receiving procainamide, a similar effect was observed, which indicates that the acute effect of a transient increase in serum procainamide induced by St. John's wort is probably mediated via modulation of intestinal P-glycoprotein (25).

Sugimoto et al. reported interactions between St. John's wort and the cholesterol-lowering drugs simvastatin and pravastatin. In a double-blind crossover study in 16 healthy male volunteers, use of St. John's wort (900 mg/day) for 14 days decreased peak serum concentrations of simvastatin hydroxyl acid, the active metabolite of simvastatin, from an average of 2.3 ng/mL in the placebo group to 1.1 ng/mL in the St. John's wort group. The simvastatin AUC was also reduced in the St. John's wort group compared to the placebo group. Simvastatin is extensively metabolized by CYP3A4 in the intestinal wall and liver, and St. John's wort induces this enzyme. By contrast, St. John's Wort did not influence plasma pravastatin concentrations (26).

5.8 Interaction between theophylline and St. John's wort

A 42-year-old female was stable on a twice-daily theophylline dose of 300 mg and had a therapeutic theophylline serum concentration of 9.2 µg/mL (therapeutic concentration 10–20 µg/mL). The patient started taking St. John's wort (0.3% hypericin, 300 mg/day) 2 months previously and was admitted to hospital, where her theophylline dose was increased to 1600 mg/day to control her clinical symptoms. Later she decided to discontinue St. John's wort and 7 days later her theophylline reached a toxic level (19.6 µg/mL) and her dose was consequently adjusted downwards (27). However, another study did not find any significant interaction between St. John's wort and theophylline (28).

5.9 Interaction between antiepileptic drugs and St. John's wort

Clinical trials have revealed that St. John's wort decreases plasma concentrations of the antiepileptic drug mephenytoin, which is primarily metabolized in the liver by isoenzymes CYP2C9 and CYP2C19 (5). Interestingly, St. John's wort does not interact with carbamazepine. Burstein et al. reported that intake of St. John's wort (900 mg/day) for 2 weeks did not alter carbamazepine pharmacokinetics. Carbamazepine is metabolized by CYP3A4, but the lack of interaction may be due to the inducing effect of carbamazepine on cytochrome P450 enzymes and therefore further induction by St. John's wort might not occur (29).

5.10 Interaction between other central nervous system (CNS)-acting drugs and St. John's wort

The benzodiazepines alprazolam and midazolam are metabolized by CYP3A4. Although short-term ingestion of St. John's wort (900 mg/day for 1–3 days) does not alter alprazolam and midazolam pharmacokinetics in healthy volunteers, long-term ingestion (900 mg/day for 2 weeks) significantly increased oral clearance of midazolam and decreased its oral bioavailability by 39.3% (30).

Depressed patients often self-medicate with St. John's wort as a herbal antidepressant. St. John's wort interacts with selective serotonin reuptake inhibitors (SSRI antidepressants) such as paroxetine, sertraline, venlafaxine and nefazodone, causing symptoms of serotonin excess. Such interactions are pharmacodynamic in nature because both St. John's wort and SSRIs have an additive effect on serotonin reuptake. This mechanism explains the occurrence of mania in a 28-year-old man taking both sertraline and St. John's wort (5). By contrast, interaction between tricyclic antidepressants and St. John's wort is pharmacokinetic in nature. Concomitant intake of St. John's wort (hypericum extract LI160) for at least 2 weeks by 12 depressed patients resulted in a decrease in $AUC_{0-12 h}$ by 22% for amitriptyline and by 41% for nortriptyline. The AUC for all hydroxylated metabolites except 10-E-hydroxynortriptyline also decreased. The mean peak amitriptyline concentration decreased from 69.8 to 54.1 ng/mL in patients taking St. John's wort. Significant reductions in peak nortriptyline concentrations were also observed in subjects taking St. John's wort (31) (therapeutic concentration amitriptyline plus nortriptyline 120–250 ng/mL). Demethylation of amitriptyline to nortriptyline is catalyzed by CYP3A4 and CYP2C19, whereas further metabolism of nortriptyline via hydroxylation at position 10 is mediated by CYP3A4 and CYP2D6.

St. John's wort probably interacts pharmacodynamically with cocaine, a drug of abuse. Saraga and Zullino reported a case of first manic episode in a patient taking St. John's wort, cocaine and alcohol (32).

5.11 Interaction between oral contraceptives and St. John's wort

Oral contraceptives are divided into two types, progestogen only and combined estrogen and progestogen. Most oral contraceptives are substrates for CYP3A4. 17-Ethynylestradiol is a major component of the oral contraceptive pill and is also used in hormonal replacement therapy in postmenopausal women. It is metabolized by hydroxylation at position 2 by CYP3A4. St. John's wort has a significant interaction with oral contraceptives. Muprhy et al. studied the interaction between St. John's wort and oral contraceptives by investigating norethindrone and ethinyl estradiol pharmacokinetics in 16 healthy women. Treatment with St. John's wort (300 mg three times a day for 28 days) resulted in a 13–15% reduction in contraceptive dose exposure. Breakthrough bleeding increased in treatment cycle, as did evidence of follicle growth and probable ovulation. The authors concluded that St. John's wort increases norethindrone and ethinylestradiol metabolism and thus interferes with contraceptive effectiveness (33).

5.12 Interaction between opioids and St. John's wort

Reduced plasma levels of methadone were observed in the presence of St. John's wort. Long-term treatment with St. John's wort (900 mg/day) for a median period of 31 days (range 14–47 days) decreased trough methadone concentrations by an average of 47% in four patients. Two patients experienced withdrawal symptoms due to reduced plasma methadone levels (34). In a crossover study of 12 healthy subjects, Nieminen et al. observed that St. John's wort significantly reduced plasma concentration of oxycodone (AUC_{0-48h} decreased by 50%). The average half-life of oxycodone decreased from 3.8 h to 3.0 h, which indicates that St. John's wort significantly increases oxycodone metabolism. This is related to CYP3A4 induction by St. John's wort because CYP3A4 is the major isoenzyme responsible for oxycodone metabolism (35).

5.13 Other clinically significant drug interactions with St. John's wort

St. John's wort accelerates both CYP3A4-catalyzed sulfoxidation and 2C19-dependent hydroxylation of omeprazole. In a study involving 12 healthy adult male volunteers receiving St. John's wort (900 mg/day) for 14 days, peak plasma concentrations of omeprazole significantly decreased after a single oral dose (20 mg). The authors concluded that there was a significant interaction between St. John's wort and omeprazole (36).

Chlorzoxazone, a centrally active muscle relaxant used in treating muscle pain, interacts with St. John's wort and the ratio of its metabolite (hydroxy chlorzoxazone) to the parent drug increases on co-administration of St. John's wort. The effect was more significant in younger people. St. John's wort did not affect the metabolism of the hypoglycemic agent tolbutamide, but significantly altered the pharmacokinetics of another hypoglycemic agent, gliclazide, and thus may cause treatment failure with this agent. In a controlled clinical trial, co-administration of St. John's wort caused a short-term clinically insignificant increase in plasma concentrations of voriconazole (an antifungal agent), followed by a prolonged significant reduction in plasma voriconazole levels (5).

5.14 Drug interactions of St. John's wort: impact on therapeutic drug monitoring

Several investigators have used commercial immunoassays to investigate interactions between St. John's wort and therapeutically monitored drugs such as theophylline, digoxin, carbamazepine, cyclosporine and tacrolimus. Fortunately, components of St. John's wort do not interfere with commercial immunoassays for therapeutic drug monitoring. It has been reported that St. John's wort does not interfere with therapeutic monitoring of 12 commonly monitored drugs using commercial immunoassays (37). However, for drugs for which immunoassays are not commercially available, various investigators have used chromatographic techniques to measure drug levels in response to administration of St. John's wort. Chromatographic techniques are usually not affected by interfering substances other than the optical isomer. However, optical isomers can also be separated by chromatographic methods using a chiral column.

As mentioned earlier, hyperforin concentration can vary widely from one St. John's wort preparation to another and drug interactions may be negligible and not of clinical concern for commercial preparations containing <1% hyperforin. Draves and Walker reported that in commercial tablets of St. John's wort, the percentage of active components varied from 31.3% to 80.2% of the claim of active ingredients based on the product label (38). Arold et al. demonstrated that low-hyperforin St. John's wort had no significant interaction with alprazolam, caffeine, tolbutamide or digoxin (39).

Although St. John's wort has clinically significant interactions with many drugs, only a few such drugs are subject to routine therapeutic monitoring. Clinically significant interactions between St. John's wort and amitriptyline, cyclosporine, tacrolimus and digoxin can be identified by observing significantly lower concentrations of these drugs in plasma (amitriptyline and digoxin) or whole blood (cyclosporine and tacrolimus). In addition, interaction between St. John's wort and warfarin can be suspected from an unexpected decrease in the International Normalized Ratio (INR). Unfortunately, for the majority of drugs that interact with St. John's wort, therapeutic drug monitoring is not usually performed, except in a few specialized academic medical centers where therapeutic drug monitoring of indinavir and other antiretroviral drugs is routine. Clinical symptoms and a lack of drug response due to reduced blood concentrations in a patient using St. Johns's wort may provide the clinician with the first indication of treatment failure. Unfortunately, the majority of patients taking herbal supplements do not inform their clinician of such use.

Another problem regarding interaction with Western drugs is that not all St. John's wort preparations show expected interactions because preparations with low hyperforin content do not induce the cytochrome P450 family of enzymes significantly enough. This could be a potentially problematic situation because another batch of preparation may have a higher hyperforin content and may result in interaction. This situation arises because herbal supplements are crude plant extract and often are not standardized. Cirak et al. reported that the hyperforin content of St. John's wort plants can vary significantly; more hyperforin accumulates as the plant grows and higher accumulation of hyperforin was observed during flowering (40). Therefore, measurement of the hyperforin content of St. John's wort may be useful to further evaluate suspected interaction between a drug and the herb. Unfortunately, there is no commercially available immunoassay for rapid determination of hyperforin content in human serum or plasma. Bauer et al. described an isocratic reverse-phase HPLC method for determination of hypericin and pseudohypericin using fluorimetric detection and hyperforin by UV detection. The limit of detection was 0.25 ng/mL for hypericin and pseudohypericin and 10 ng/mL for hyperforin in human plasma (41). Recently, Piovan et al. described a flow injection analysis-electrospray ionization-mass spectrometric (FIA-ESI-MS) method for rapid screening of St. John's wort extracts. This method can semi-quantitatively estimate hyperforin and other active components of St. John's wort, such as chlorogenic acid, rutin, quercetin, quercitrin and hypericin (42). However, these methods require sophisticated analytical techniques and experienced personnel. Only large tertiary care facilities, reference laboratories and clinical laboratories in major academic medical centers have such facilities.

5.15 Conclusions

St. John's wort may cause both pharmacokinetic and pharmacodynamic interactions with several Western drugs. Pharmacokinetic interactions with St. John's wort are more common and may lead to treatment failure owing to lower blood levels of many drugs caused by increased drug metabolism due to hyperforin induction of drug-metabolizing liver enzymes CYP3A4, CYP2E1 and CYP2C19. Modulation of P-glycoprotein by hypericin is the mechanism by which St. John's wort reduces plasma levels of certain drugs that are not subject to major hepatic metabolism, such as digoxin. Pharmacodynamic interactions with St. John's wort have been reported for drugs that modulate serotonin uptake. Clinically significant pharmacodynamic interactions causing serotonin syndrome have been reported when St. John's wort is combined with SSRIs or serotonin receptor agonists.

Other than a few drugs that are subject to routine therapeutic drug monitoring, identification of a clinically significant interaction between St. John's wort and a drug is based on clinical diagnosis because there is no easily available laboratory test to measure levels of many drugs that interact with St. John's wort. Therefore, it is important for clinicians to question their patients regarding use of herbal supplements to identify such interactions. Fortunately, after discontinuation of St. John's wort, the drug-herb interaction resolves within a few days to a week. In addition, where interaction between St. John's wort and a drug may lead to a potentially life-threatening situation, such as in organ transplant recipients and patients attending a warfarin clinic, it is important for healthcare professionals to educate their patients regarding the danger of drug-herb interactions. It is also advisable for AIDS patients on antiretroviral therapy to avoid any herbal supplements, including St. John's wort. The use of all herbal supplements must be discontinued at least 3 weeks prior to elective surgery to avoid potentially dangerous interactions between a herbal supplement and an anesthetic agent. Moreover, garlic (see color plate), ginger (see color plate), ginkgo (see color plate) and related supplements must also be avoided to prevent any unusual bleeding episodes during surgery.

References

1. Karilova, M., Treichel, U., Malago, M., Frilling, A., Gerkin, G., Broelsch, C.E. (2000) Interaction of Hypericum perforatum (St. John's wort) with cyclosporine A metabolism in a patient after liver transplant. J. Hepatol. 33:853–855.
2. Landrum-Michalets, E. (1998) Update: clinically significant cytochrome P450 drug interactions. Pharmacotherapy 18:84–112.
3. Wentworth, J.M., Agostini, M., Love, J., Schwabe, J.W., Chatterjee, V.K. (2000) St. John's wort, a herbal antidepressant, activates the steroid X receptor. J. Endocrinol. 166:R11–16.
4. Raffa, R. (1998) Screen of receptor and uptake site activity of hypericin components of St. John's wort reveals σ receptor binding. Life Sci. 62:265–270.

5. Izzo, A.A., Ernst, E. (2009) Interaction between herbal medicines and prescribed drugs: an updated systematic review. Drugs 69:1777–1798.
6. Choudhuri, S., Valerio, L.G., Jr. (2005) Usefulness of studies on the molecular mechanism of action of herbals/botanicals: the case of St. John's wort. J. Biochem. Mol. Toxicol. 19:1–11.
7. Godtel-Armbrust, U., Metzger, A., Kroll, U., Kelber, O., Wojnowski, L. (2007) Variability in PXR-mediated induction of CYP3A4 by commercial preparations and dry extracts of St. John's wort. Naunyn Schmiedebergs Arch. Pharamcol. 375:372–382.
8. Madabushi, R., Frank, B., Drewelow, B., Derendorf, H., Butterweck, V. (2006) Hyperforin in St. John's wort drug interactions. Eur. J. Pharamcol. 62:225–233.
9. Barone, G.W., Gurley, B.J., Ketel, B.L., Abul-Ezz, S.R. (2001) Herbal supplements; a potential for drug interactions in transplant recipients. Transplantation 71:239–241.
10. Alscher, D.M., Klotz, U. (2003) Drug interaction of herbal tea containing St. John's wort with cyclosporine. Transpl. Int. 16:543–544.
11. Mai, I., Bauer, S., Perloff, E.S., Johne, A., Uehleke, B., Frank, B. et al. (2004) Hyperforin content determines the magnitude of the St. John's wort-cyclosporine drug interaction. Clin. Pharmacol. 76:330–340.
12. Hebert, M.F., Park, J.M., Chen, Y.L., Akhtar, S., Larson, A.M. (2004) Effects of St. John's wort (Hypercium perforatum) on tacrolimus pharmacokinetics in healthy volunteers. Clin. Pharmacol. 44:89–94.
13. Bolley, R., Zulke, C., Kammerl, M., Fischereder, M., Kramer, B.K. (2002) Tracrolimus-induced nephrotoxicity unmasked by induction of CYP3A4 system with St. John's wort. Transplantation 73:1009.
14. Mai, I., Stormer, E., Bauer, S., Kruger, H., Budde, K., Roots, I. (2003) Impact of St. John's wort treatment on the pharmacokinetics of tacrolimus and mycophenolic acid in renal transplant patients. Nephrol. Dial. Transplant 18:819–822.
15. James, J.S. (2000) St. John's wort warning: do not combine with protease inhibitors, NNRTIs. AIDS Treat. News 18(337):3–5.
16. van den Bout-van den Beukel, C.J., Koopmans, P.P., van der Ven, A.J., De Smet, P.A., Burtger, D.M. (2006) Possible drug metabolism interactions of medicinal herbs with antiretroviral agents. Drug Metab. Rev. 38:477–514.
17. Piscitelli, S.C., Burstein, A.H., Chaitt, D., Alfaro, R.M., Fallon, J. (2000) Indinavir concentrations and St. John's wort. Lancet 355:547–548.
18. Busti, A.J., Hall, R.J., Margolis, D.M. (2004) Atazanavir for the treatment of human immunodeficiency virus infection. Pharmacotherapy 24:1732–1747.
19. Cvetkovic, R.S., Goa, K.L. (2003) Lopinavir/ritonavir: a review of its use in the management of HIV infection. Drugs 63:769–802.
20. De Maat, M.M., Hoetelmans, R.M., Mathot, Gorp, E.C., Meenhorst, P.L., Mulder, J.W. et al. (2001) Drug interaction between St. John's wort and nevirapine. AIDS 15:420–421.
21. Smith, P. (2004) The influence of St. John's wort on the pharmacokinetics and protein binding of imatinib mesylate. Pharmacotherapy 24:1508–1514.
22. Mathijssen, R.H., Verweij, J., de Bruijn, P., Loos, W.J., Sparreboom, A. (2002) Effects of St. John's wort on irinotecan metabolism. J. Natl. Cancer Inst. 94:1247–1249.
23. Johne, A., Brockmoller, J., Bauer, S., Maurer, A., Langheinrich, M., Roots, I. (1999) Pharmacokinetic interaction of digoxin with an herbal extract from St John's wort (Hypericum perforatum). Clin. Pharmacol. Ther. 66:338–345.
24. Durr, D., Stieger, B., Kullak-Ublick, G.A., Rentsch, K.M., Steinert, H.C., Meier, P.J. et al. (2000) St. John's wort induces intestinal P-glycoprotein/MDR1 and intestinal and hepatic CYP3A4. Clin. Pharmacol. Ther. 68:598–604.
25. Dasgupta, A., Hovanetz, N., Olsen, M., Wells, A., Actor, J.K. (2007) Drug-herb interaction: effect of St. John's wort on bioavailability and metabolism of procainamide in mice. Arch. Pathol. Lab. Med. 131:1094–1098.

26. Sugimoto, K., Ohmori, M., Tsuruoka, S., Nishiki, K., Kawaguchi, A., Harada, K. et al. (2001) Different effects of St. John's wort on the pharmacokinetics of simvastatin and pravastatin. Clin. Pharmacol. Ther. 70:518–524.

27. Nebel, A., Schneider, B.J., Kroll, D.J. (1999) Potential metabolic interaction between St. John's wort and theophylline. Ann. Pharmacother. 33:502.

28. Morimoto, T., Kotegawa, T., Tsutsumi, K., Ohtani, Y., Imai, H., Nakano, S. (2004) Effect of St. John's wort on the pharmacokinetics of theophylline in healthy volunteers. J. Clin. Pharmacol. 44:95–101.

29. Burstein, A.H., Horton, R.L., Dunn, T., Alfaro, R.M., Piscitelli, S.C., Theodore, W. (2000) Lack of effect of St. John's wort on carbamazepine pharmacokinetics in healthy volunteers. Clin. Pharmacol. Ther. 68:605–612.

30. Wang, Z., Gorski, J.C., Hamman, M.A., Huang, S.M., Lesko, L.J., Hall, S.D. (2001) The effects of St. John's wort (Hypericum perforatum) on human cytochrome P450 activity. Clin. Pharmacol. Ther. 70:317–326.

31. Johne, A., Schmider, J., Brockmoller, J., Stadelman, A.M., et al. (2002) Decreased plasma levels of amitriptyline and its metabolites on comedication with an extract from St. John's wort (Hypericum perforatum). J. Clin. Psychopharmacol. 22:46–54.

32. Saraga, M., Zullino, D.F. (2005) St. John's wort, corticosteroids, cocaine, alcohol and a first manic episode. Praxis 94:987–989 (in French).

33. Murphy, P.A., Kern, S.E., Stanczyk, F.Z., Westhoff, C.L. (2005) Interaction of St. John's wort with oral contraceptives: effects on the pharmacokinetics of norethindrone and ethinyl estradiol, ovarian activity and breakthrough bleeding. Contraception 71:402–408.

34. Eich-Hochli, D., Oppliger, R., Golay, K.P., Baumann, P., Eap, C.B. (2003) Methadone maintenance treatment and St. John's wort – a case study. Pharmacopsychiatry 36:35–37.

35. Nieminen, T.H., Hagelberg, N.M., Saari, T.I., Neuvonen, M., Laine, K., Neuvonen, P.J. et al. (2010) St. John's wort greatly reduces the concentrations of oral oxycodone. Eur. J. Pain 14:854–859.

36. Wang, L.S., Zhou, G., Zhu, B., Wu, J., Wang, J.G., Abd El-Aty, A.M. et al. (2004) St. John's wort induces both cytochrome P450 3A4-catalyzed sulfoxidation and 2C19-dependent hydroxylation of omeprazole. Clin. Pharmacol. Ther. 75:191–197.

37. Dasgupta, A., Tso, G., Szelei-Stevens, K.I. (2006) St. John's wort does not interfere with therapeutic drug monitoring of 12 commonly monitored drugs using immunoassays. J. Clin. Lab. Anal. 20:62–67.

38. Draves, A.H., Walker, S.E. (2003) Analysis of hypericin and pseudohypericin content of commercially available St. John's wort preparations. Can. J. Clin. Pharmacol. 10:114–118.

39. Arold, G., Donath, F., Maurer, A., Diefenbach, K., Bauer, S., Henneicke-von Zepelin, H.H. et al. (2005) No relevant interaction with alprazolam, caffeine, tolbutamide and digoxin by treatment with a low-hyperforin St. John's wort extract. Planta Med. 71:331–337.

40. Ciark, C., Radusiene, J., Janulis, V., Ivanauskas, L. (2008) Pseudohypericin and hyperforin in Hypericum perforatum from Northern Turkey: variation among populations, plant parts and phenological stages. J. Int. Plant Biol. 50:575–580.

41. Bauer, S., Stromer, E., Graubaum, H.J., Roots, I. (2010) Determination of hyperforin, hypercin and pseudohypericin in human plasma using high performance liquid chromatography analysis with fluorescence and ultraviolet detection. J. Chromatogr. B Biomed. Sci. Appl. 765:29–35.

42. Piovan, A., Filippini, R., Caniato, R. (2010) A semi-quantitative FIA-ESI-MS method for the rapid screening of Hypericum perforatum crude extract. Nat. Prod. Commun. 5:431–434.

6 Herbs to avoid with warfarin therapy

6.1 Introduction

Warfarin (common brand name Coumadin) is an anticoagulant used to prevent thrombosis and embolism. Warfarin was initially marketed as a poison to kill rats and mice, but was later used as a drug for human use. Warfarin is a synthetic derivative of dicoumarol. Dicoumarol is derived from coumarin, a naturally occurring product found in some plant species. Approved for clinical use in the 1950s, this drug is still commonly used today as an anticoagulant. Warfarin is a racemic mixture of two active enantiomers, but the S-form is approximately five times more active than R-warfarin. Warfarin inhibits vitamin K-dependent synthesis of biologically active calcium-dependent clotting factors such as II, VII, IX and X. Other regulatory proteins, such as protein C, protein S and protein Z, that play important roles in the clotting mechanism are also inhibited. Warfarin is slower-acting than the anticoagulant heparin, but unlike heparin, which must be administered by injection, warfarin can be given orally. Warfarin has a long half-life and can be administered as a once-daily dose. Warfarin activity is determined by genetic factors. Polymorphisms of two genes (*VKORC1* and *CYP2C9*) are of particular importance.

Warfarin is known to interact with many Western drugs. In addition, certain herbal supplements potentiate warfarin action, whereas other supplements may reduce its efficacy. Certain herbal supplements may also interfere with bioavailability of orally administered warfarin. In this chapter, important interactions between warfarin and herbal supplements are discussed.

6.2 Interaction between warfarin and herbal supplements

Although warfarin concentrations are not usually measured in clinical therapy, warfarin therapy is closely monitored in clinical laboratories by measuring the international normalization ratio (INR). In general, the initial INR target is 2–3 during warfarin therapy, but a different target may be set, depending on the clinical condition and a particular patient. Herbal supplements may potentiate or reduce the efficacy of warfarin therapy. Certain herbal supplements may also interfere with warfarin absorption from the gut and thus affect warfarin bioavailability and hence its efficacy. Herbal supplements that increase or reduce the effects of warfarin are listed in ▶Tab. 6.1. The chemical structure of warfarin is given in ▶Fig. 6.1.

6.3 Herbal supplements that increase warfarin efficacy

Many herbal supplements are known to potentiate the effect of warfarin and may lead to excessive anticoagulation, which is a potentially dangerous clinical situation. In such

Tab. 6.1: Herbal products and food that interact with warfarin therapy

Herbal supplement	Effect on warfarin
Boldo	Increases the efficacy
Borage	Increases the efficacy
DanShen	Increases the efficacy
Dong quai	Increases the efficacy
Devil's claw	Increases the efficacy
Fenugreek	Increases the efficacy
Feverfew	Increases the efficacy
Grape seed	Increases the efficacy
Garlic	Increases the efficacy
Ginkgo biloba	Increases the efficacy
Papaya	Increases the efficacy
Mango	Increases the efficacy
Fish oil supplements	Increases the efficacy
Evening primrose oil	Increases the efficacy
Horse chestnut	Increases the efficacy
Vitamin E	Increases the efficacy
Royal jelly	Increases the efficacy
Saw palmetto	Increases the efficacy
St. John's wort	Reduces the efficacy
Green tea	Reduces the efficacy
Milk Thistle	Reduces the efficacy
Ginseng	Reduces the efficacy
Goldenseal	Reduces the efficacy
Soy milk	Reduces the efficacy

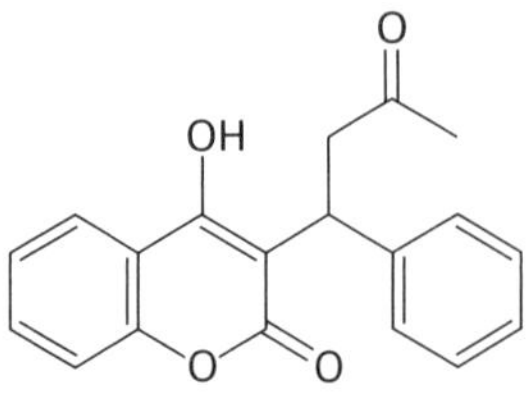

Fig. 6.1: Chemical structure of warfarin.

cases an increase in INR with no change in dosage may be an early indication of such warfarin-herb interactions. Approximately 30% of patients who take warfarin also ingest herbal supplements and therefore are at a particular risk of warfarin-herb interaction. In addition, there are numerous patient-specific influences that may be responsible for a particular observation. Most reported interactions between warfarin and various herbal supplements are based on case studies and there are only a few systematic studies involving a large number of subjects (1). In general it is assumed that angelica root, anise, asafoetida, bogbean, borage seed oil, bromelain, capsicum, chamomile, clove, fenugreek, feverfew, garlic (see color plate), ginger (see color plate), ginkgo biloba (see color plate), horse chestnut, licorice root, meadowsweet, passionflower, red clover, turmeric extract and willow bark potentially increase the effectiveness of warfarin and thus increase the risk of bleeding in patients taking warfarin and one of these supplements (2). Shalansky et al. studied 171 adult patients taking warfarin and observed that 73 patients (43%) used at least one herbal supplement with a known interaction with warfarin, indicating the severity of clinical problems associated with interactions between warfarin and herbs in such patients. Herbal supplements associated with self-reported bleeding episodes include cayenne, ginger, willow bark and coenzyme Q10. In addition to warfarin-herb interactions, other risk factors associated with increased bleeding include a high INR target (2.5–3.5), diarrhea, acetaminophen use, increased alcohol consumption and increased age. Use of ginger supplements and coenzyme Q10 were also independently associated with increased risk of bleeding in patients taking warfarin (3).

In a case study a significant interaction was observed between warfarin and boldo-fenugreek. The patient was taking warfarin for atrial fibrillation. During treatment an unexpected increase in INR (range varying with condition, usually 2–3) was observed and the patient admitted to taking herbal supplements, including boldo and fenugreek. When she stopped taking these supplements her INR returned to normal after 1 week. The herb-warfarin interaction was observed again when the patient took the supplements again. The authors concluded that herbal supplements such as boldo and fenugreek may increase bleeding time in a patient taking warfarin (4). Interaction between warfarin and DanShen has been reported. DanShen is commonly used for atherosclerosis-related disorders. DanShen can affect hemostasis in multiple ways, including inhibition of platelet aggregation, exertion of antithrombin III-like activity and promotion of fibrinolytic activity. Experiments in a rat model indicated that DanShen increases the absorption rate constant, area under the curve (AUC), maximum plasma concentration and elimination half-life of warfarin. DanShen also decreased the clearance and volume of distribution of both R- and S-warfarin. Thus, DanShen exaggerates the pharmacological action of warfarin. There are three case reports of gross overcoagulation and bleeding episodes in patients on chronic warfarin therapy who also took DanShen. Because of both pharmacokinetic and pharmacodynamic interactions between DanShen and warfarin, patients taking warfarin must avoid DanShen (5). A 46-year-old woman on warfarin therapy experienced overcoagulation after starting to take dong quai to relieve menopause symptoms. Her INR returned to normal after she stopped taking dong quai. Women taking warfarin must refrain from using dong quai to obtain relief from menopause discomfort because dong quai contains coumarin derivatives (6).

Garlic supplements may reduce the risk of atherosclerosis, most probably because of its antiplatelet activity. Garlic supplements may increase the risk of bleeding without warfarin therapy. Garlic may enhance the pharmacological effect of warfarin and another anticoagulant, fluindione (7). Ginkgo biloba is known to have antiplatelet activity. In one case, a patient on warfarin therapy for 5 years experienced a left parietal hemorrhage 2 months after initiating ginkgo biloba therapy. This was attributed to the additive effect of ginkgo and warfarin in anticoagulation (8). Intracerebral hemorrhage associated with long-term ginkgo therapy has been reported even in the absence of anticoagulation therapy (9).

Fish oil contains omega 3 fatty acids and can inhibit platelet aggregation. Potentiation of warfarin action by fish oil is well documented in the literature. Buckley et al. reported a case of a 67-year-old woman who was taking warfarin for 1.5 years owing to recurrent transient ischemic attacks. Her medical history indicated that she was suffering from hypothyroidism, hyperlipidemia, osteopenia, hypertension and coronary artery disease. Her INR was checked regularly during warfarin therapy and was within the acceptable range when she was taking 1.5 mg of warfarin per day. When the patient started taking a fish oil supplement her INR increased from 2.8 to 4.3 within 1 month. Her INR decreased to 1.6 within a week after she reduced her fish oil intake and her warfarin dose was adjusted back to the original level. Fish oil contains omega 3 fatty acids such as eicosapentaenoic acid and docosahexaenoic acid that can reduce the supply of thromboxane A2 supplies within platelets and may also reduce factor VIII levels. The authors concluded that patients undergoing warfarin therapy should be educated regarding drug-herb interactions and must be monitored for such interactions. Pharmacists can play an important role in identifying such interactions by asking patients on warfarin about their use of herbal supplements (10). In another case report, an elderly patient developed a subdural hematoma after a minor fall that required craniotomy. Development of a subdural hematoma after such a minor fall was probably related to high-dose consumption of fish oil (6 g/day) along with warfarin and aspirin (11). Evening primrose oil also has antiplatelet activity and may potentiate the pharmacological action of warfarin.

Royal jelly, a secretion by honey bees to nurture larvae, is used as a natural supplement. Royal jelly may potentiate the action of warfarin. An 87-year-old African-American man presented to an internal medicine clinic for a routine anticoagulation check during warfarin therapy. His medical history included a stage IV-A follicular non-Hodgkin's lymphoma, atrial fibrillation and hypertension. His long-term drug therapy consisted of warfarin, felodipine, lisinopril-hydrochlorothiazide, controlled-release diltiazem, potassium chloride and oxycodone. His INR ranged from 1.9 to 2.4. On admission to the hospital his INR was 6.88, which subsequently increased to 7.29 during his hospital stay. On further investigation, the patient admitted to taking a royal jelly supplement 1 week before his hospital admission. The most probable explanation for his abnormally elevated INR and subsequent bleeding was potentiation of warfarin action by royal jelly (12).

6.4 Herbal supplements that reduce warfarin efficacy

St. John's wort (see color plate) (see chapter 5) is a herbal antidepressant that is popular throughout the world. St. John's wort interacts with many drugs and reduces their efficacy due to increased clearance. Using 12 healthy volunteers, Jiang et al. demonstrated that St. John's wort induced clearance of both *R*- and *S*-warfarin, which in turn resulted in a significant reduction in pharmacological action of racemic warfarin. However, co-administration of warfarin and ginseng (see color plate) did not affect the pharmacokinetics or pharmacodynamics of either *R*- or *S*-warfarin (13). However, in a randomized controlled trial in 20 healthy subjects, Yuan et al., observed that the peak INR decreased significantly after 2 weeks of American ginseng administration in subjects also receiving warfarin compared to the placebo group. The peak plasma concentration and AUC for warfarin were also significantly reduced in the ginseng compared with the placebo group. The authors concluded that American ginseng reduces the anticoagulant effect of warfarin and physicians should ask patients about ginseng use when prescribing warfarin (14).

Green tea consumption is associated with reduced pharmacological action of warfarin. A 44-year-old Caucasian man, who was on warfarin for thromboembolic prophylaxis secondary to St. Jude mechanical valve replacement in the aortic position, had an INR of 3.79 at a clinic visit. His INR was 1.37 after 22 days and further decreased to 1.14 after 1 month. It was subsequently discovered that the patient was drinking half to one gallon of green tea per day for approximately 1 week before his INR decreased to 1.37. On discontinuation of green tea, his INR increased to 2.55. Warfarin has an anticoagulant effect by inhibiting vitamin K, depending on clotting factors such as factors II, VII, IX and X, exogenous administration of vitamin K antagonizes the effect of warfarin, as observed by a reduction in INR. Green tea contains significant amounts of vitamin K and thus reduces the efficacy of warfarin (15). Consumption of soy milk may also reduce the efficacy of warfarin. A 70-year-old Caucasian man who was stable on warfarin therapy developed sub-therapeutic INR values after ingesting soy protein in the form of soy milk. His INR returned to normal 2 weeks after discontinuing consumption of soy milk (16).

6.5 Fruits, vegetables, vitamins and mineral supplements, and other agents that interact with warfarin

One publication reported on 13 male patients whose INR values increased significantly after consumption of mango fruit. After the possible cause of the INR increase was identified, the patients were instructed to stop consuming mango and their INR values decreased to therapeutic levels. Although the exact mechanism of this fruit-warfarin interaction is not completely understood, mango is a rich source of vitamin A, which inhibits the CYP2C19 enzyme responsible for warfarin metabolism. Modest inhibition of this drug-metabolizing enzyme may lead to great warfarin efficacy (17). Papaya may also interact with warfarin by increasing its pharmacological action. There is a case report of increased INR in a patient on warfarin after addition of papaya extract to his prescribed medication. Papaya is contraindicated for warfarin because it may damage the mucous membrane of the gastrointestinal tract and can result in increased bleeding (18).

Like green tea, leafy vegetables are also a rich source of vitamin K. The vitamin K content of turnip greens (650 μg/100 g of vegetable) and broccoli (200 μg/100 g of vegetable) is very high and intake of these vegetables on a regular basis may reduce the efficacy of warfarin (19). Very large amounts of vitamin K from a single meal of vegetables (400 g of vegetables delivering 700–1500 mg of vitamin K) can lead to measurable INR changes, but an occasional typical serving of <100 g of vegetables may have an insignificant effect on INR in patients on warfarin therapy (20). The vitamin content of some common vegetables is listed in ▶Tab. 6.2.

Coenzyme Q10 is structurally similar to vitamin K and may affect warfarin efficacy. There are several case reports of potential interaction of coenzyme Q10 with warfarin. A 72-year-old female patient being treated with warfarin showed less responsiveness to the drug than previously observed. Her medical history revealed that she had started taking coenzyme Q10. When she stopped taking the supplement, her responsiveness to warfarin returned to the previously observed level. The authors concluded that because coenzyme Q10 is structurally similar to vitamin K, co-administration of coenzyme Q10 may reduce the efficacy of warfarin (21). However, a randomized, double-blind, placebo-controlled clinical trial involving 14 women and 10 men revealed no effect of coenzyme Q10 on the pharmacological action of warfarin (22).

Iron, magnesium and zinc may bind to warfarin and can thus potentially decrease its bioavailability by reducing absorption from the gut. Therefore, such mineral supplements must be consumed at least 2 h before oral administration of warfarin (23).

Alcohol consumption may also affect the efficacy of warfarin therapy. A 58-year-old Caucasian male was on long-term warfarin therapy to prevent ischemic stroke. His INR was stable but increased when he started to consume low amounts of beer. When he stopped drinking beer, his INR returned to normal. This case illustrates a potential for low-dose beer consumption leading to elevated INR. This may be due to binding of warfarin to serum proteins and modulation of alcohol metabolism (24). There is one case

Tab. 6.2: Vitamin K content of some vegetables

Vegetable	Typical serving (g)	Vitamin K content (µg)
Kale	130	1146.6
Spinach (frozen)	190	1027.3
Turnip greens	164	851.0
Brussel sprout	155	299.0
Broccoli	156	220.0
Spring onion (raw)	100	207.0
Lettuce	163	166.7
Cabbage	170	57.8
Coleslaw	99	56.4
Peas	160	40.0
Cauliflower	100	17.1

Source: United States Department of Agriculture website: http://www.nal.usda.gov/fnic/foodcomp/Data/SR17, accessed July 10, 2010.

report of increased INR in a patient after smoking cessation. A 58-year-old male smoker taking warfarin at a stable dose was admitted to hospital with a diagnosis of bacterial meningitis. After admission, he decided to stop smoking. He was maintained on the same warfarin dose but after discharge his INR increased substantially. Eventually his warfarin dose was decreased by 23%. This case illustrates the potential for an interaction between warfarin and cigarette smoking (25). Menthol cough drops may also reduce the efficacy of warfarin. A 46-year-old African American man taking warfarin for a venous thromboembolism experienced a decrease in INR from 2.3 to 1.6. The patient reported that he was taking eight to ten menthol cough drops. Five days after discontinuing the cough drops, his INR increased to 2.9. Menthol probably alters warfarin metabolism by modulating the cytochrome liver enzyme involved. The authors also noted that to their knowledge this was the second case report describing an interaction between menthol and warfarin (26).

6.6 Conclusions

In addition to various herbal supplements, vitamins and minerals, various drugs also interact significantly with warfarin therapy. However, a discussion on warfarin-drug interactions is beyond the scope of this book. Fortunately, physicians and pharmacists are aware of drug-warfarin interactions and can carefully manage treatment regimes for their patients. Unfortunately, many patients consider herbal supplements safe and do not report their use of various herbal supplements to their doctor. As discussed in this chapter, many herbs can significantly interact with warfarin, either increasing or decreasing its efficacy. Because both undercoagulation and overcoagulation are dangerous in patients on warfarin therapy, tight INR must be maintained in these patients. Physicians, pharmacist, nurses and other healthcare providers must educate their patients so that they discuss the use of any herbal supplements beforehand to avoid an adverse warfarin-herb interaction.

References

1. Wittkowsky, A.K. (2008) Dietary supplements, herbs and oral anticoagulants: the nature of the evidence. J. Thromb. Thrombolysis 25:72–77.
2. Heck, A.M., DeWitt, B.A., Lukes, A.L. (2000) Potential interactions between alternative therapies and warfarin. Am. J. Health Syst. Pharm. 57:1228–1230.
3. Shalansky, S., Lynd, L., Richardson, K., Ingaszewski, A., Kerr, C. (2007) Risk of warfarin related bleeding events and supratherapeutic international normalized ratios associated with complementary and alternative medicines: a longitudinal analysis. Pharmacotherapy 27:1237–1247.
4. Lambert, J.P., Cormier, J. (2001) Potential interaction between warfarin and boldo-fenugreek. Pharmacotherapy 21:509–512.
5. Chan, T.Y. (2001) Interaction between warfarin and DanShen (Salvia miltiorrhiza). Ann. Pharmacother. 35:501–504.
6. Page, R.L., Lawrence, J.D. (1999) Potentiation of warfarin by dong quai. Pharmacotherapy 19:870–876.
7. Borrelli, F., Capasso, R., Izzo, A.A. (2007) Garlic (Allium sativum L.): adverse effects and drug interactions in humans. Mol. Nutr. Food Res. 51:1386–1397.

8. Matthews, M.K. (1998) Association of Ginkgo biloba with intracerebral hemorrhage. Neurology 50:1933–1934.
9. Rowing, J., Lewis, S.L. (1996) Spontaneous bilateral subdural hematomas associated with chronic ginkgo biloba ingestion. Neurology 46:1775–1776.
10. Buckley, M.S., Goff, A.D., Knapp, W.E. (2004) Fish oil interaction with warfarin. Ann. Pharmacother. 38:50–52.
11. McClaskey, E.M., Michallets, E.L. (2007) Subdural hematoma after a fall in an elderly patient taking high-dose omega-3 fatty acids with warfarin and aspirin: case report and review of the literature. Pharmacotherapy 27:152–160.
12. Lee, N.J., Fermo, J.D. (2006) Warfarin and royal jelly interaction. Pharmacotherapy 26:583–586.
13. Jiang, X., Williams, K.M., Liauw, W.S., Ammit, A.J., Roufogalis, B.D., Duke, C.C. et al. (2004) Effect of St. John's wort and ginseng on the pharmacokinetics and pharmacodynamics of warfarin in healthy subjects. Br. J. Clin. Pharmacol. 57:592–599.
14. Yuan, C.S., Wei, G., Dey, L., Karrison, T., Nahlik, L., Maleckar, S. et al. (2004) American ginseng reduces warfarin's effect in healthy patients: a randomized controlled trial. Ann. Intern. Med. 141:23–27.
15. Taylor, J.R., Wilt, V.M. (1999) Probable antagonism of warfarin by green tea. Ann. Pharmacother. 33:426–428.
16. Cambria-Kiley, J.A. (2002) Effect of soy milk on warfarin efficacy. Ann. Pharmacother. 36:1893–1896.
17. Monterrey-Rodriguez, J. (2002) Interaction between warfarin and mango fruit. Ann. Pharmacother. 36:940–941.
18. Izzo, A.I., Di Carlo, G., Borrelli, F., Ernst, E. (2005) Cardiovascular pharmacotherapy and herbal remedies: the risk of drug interaction. Int. J. Cardiol. 98:1–14.
19. Cheng, T.O. (2008) Not only green tea, but also green leafy vegetables inhibit warfarin. Int. J. Cardiol. 125:101.
20. Johnson, M.A. (2005) Influence of vitamin K on anticoagulant therapy depends on vitamin K status and the source and chemical form of vitamin K. Nutr. Rev. 63:91–97.
21. Lando, C., Almdal, T.P. (1998) Interaction between warfarin and coenzyme Q10. Ugeskr. Laeger 160:3226–3227 (in Danish).
22. Engelsen, J., Nielsen, J.D., Hansen, K.F. (2003) Effect of coenzyme Q10 and ginkgo biloba on warfarin dosage in patients on long-term warfarin therapy. A randomized, double-blind, placebo-controlled cross-over trial. Ugeskr. Laeger 165:18698-1871 (in Danish).
23. Pinto, J.T. (1991) The pharmacokinetic and pharmacodynamic interactions of foods and drugs. Top. Clin. Nutr. 6:14–33.
24. Havrda, D.E., Mai, T., Chonlahan, J. (2005) Enhanced antithrombotic effect of warfarin associated with low dose alcohol consumption. Pharmacotherapy 25:303–307.
25. Evans, M., Lewis, G.M. (2005) Increase in international normalization ratio after smoking cessation in a patient receiving warfarin. Pharmacotherapy 25:1656–1659.
26. Coderre, K., Faria, C., Dyer, E. (2010) Probable warfarin interaction with menthol cough drops. Pharmacotherapy 30:110.

7 Interaction of ginseng, ginkgo, garlic and ginger supplements with various drugs

7.1 Introduction

Herbal remedies are widely used by the general population not only in developing countries, but also in developed countries. Patients suffering from chronic disease use complementary and alternative medicines including herbal supplements more frequently than the general population. Hasan et al. reported that out of 321 patients they interviewed, 205 (63.8%) used complementary and alternative medicines including various herbal supplements. Among these patients, a significant number (35.5%) were suffering from diabetes mellitus. The supplements most commonly used by these patients were vitamins (48.2%), herbal supplements (26.4%) and ginseng (see color plate) (4.7%) (1). In another study, the authors found that the most popular herbal supplements used by the general population in the USA were echinacea (see color plate), garlic (see color plate), ginseng (see color plate) and ginger (see color plate) (2). Ginkgo biloba (see color plate) is another popular herbal supplement used by older people to sharpen their memory. In this chapter, drug interactions involving ginseng, ginkgo, garlic and ginger supplements are addressed. Important drug interactions are listed in ▶Tab. 7.1.

Tab. 7.1: Interaction of ginseng, ginkgo, garlic and ginger supplements with various drugs

Herbal supplement	Interacting drug	Comments
Asian ginseng	Alcohol	Increased clearance
Asian ginseng	Phenelzine	Coma
Asian ginseng	Warfarin	Reduced effect of warfarin
Ginkgo biloba	Aspirin	Increased risk of bleeding
Ginkgo biloba	Diclofenac	Increased risk of bleeding
Ginkgo biloba	Warfarin	Increased INR
Ginkgo biloba	Trazodone	Coma
Ginkgo biloba	Omeprazole	Decreased plasma level
Garlic	Saquinavir	Decreased plasma level
Garlic	Warfarin	Increased INR
Ginger	Warfarin	Increased INR
Ginger	Nifedipine	Potentiation of the antiplatelet effect of nifedipine

7.2 Ginseng: efficacy and toxicity

Ginseng is a herbal product that is widely used in China, other Asian countries, the USA and other developed countries. For thousands of years, the Chinese have used ginseng as a tonic and in emergency medicine to rescue dying patients. Ginseng means 'essence of man' in Chinese. The Chinese ginseng that grows in Manchuria is *Panax ginseng*. However, the ginseng that grows in North America is *Panax quinquefolius*. The most common preparation is ginseng root. In the American market, ginseng is sold as tablets or capsules. Dried ginseng root or its extract is also available. Ginseng is promoted as a tonic capable of invigorating the user physically, mentally and sexually. It is also used to deal with stress because of its calming effect. Ginseng may also be effective in the treatment of mild hyperglycemia and is used for treating anemia. In Germany it is also indicated to combat a lack of energy. Ginseng contains saponins known as panaxosides or ginsenosides. It also contains antioxidants such as maltol, salicylic acid and vanillic acid. Several peptides, polysaccharides, fatty acids, cholesterol ester transfer protein inhibitors and vitamins are also found in ginseng. At least 28 ginsenosides have been isolated and characterized (3). Interestingly, some ginsenosides have biological effects in direct opposition to effects produced by others. For example, ginsenoside R_{b1} has a suppressive effect on the central nervous system (CNS), whereas ginsenoside R_{g1} has a stimulatory effect on the CNS. Some ginsenosides can reduce stress and can also reduce blood sugar.

Siberian ginseng is different from Asian ginseng because it does not contain *Panax*-type ginsenosides. Siberian ginseng is prepared from the root of *Eleutherococcus senticosus* and is also imported from China. Hikino et al. isolated seven glycans (eleutherans) from a crude extract of Siberian ginseng. This aqueous extract decreased plasma sugar levels in mice (4).

In 1979, the term ginseng abuse syndrome was coined as a result of a study on 133 people who took ginseng for 1 month. Most subjects experienced CNS stimulation. The effect of ginseng on mood seems to be dose-dependent. At a dose of <15 g/day, subjects experienced depersonalization and confusion. At a dose >15 g/day, some subjects experienced depression. Fourteen patients experienced ginseng abuse syndrome, which is characterized by symptoms of hypertension, nervousness, sleeplessness, skin eruption and morning diarrhea (5). An episode of Steven-Johnson syndrome was reported in a 27-year-old man following ingestion of ginseng at a dose of two pills for 3 days. The patient recovered after 30 days (6). Vaginal bleeding has been reported in cases related to ginseng use. Interestingly, one patient had undergone a hysterectomy 14 years earlier (7).

7.3 Ginkgo biloba: efficacy and toxicity

Ginkgo biloba is prepared from dried leaves of the ginkgo tree by organic extraction (acetone/water). After the solvent is removed, the extract is dried and standardized. Most commercial dosage forms contain 40 mg of this extract. The ginkgo tree was brought from China to Europe. Ginkgo fruits and seeds have been used in China since 2800 BC. Ginkgo biloba is sold in the USA as a dietary supplement to improve blood flow in the brain and peripheral circulation. It is used mainly to sharpen mental focus and to improve diabetes-related circulatory disorders. It is also used to cure impotence and vertigo. The German Commission E approved the use of ginkgo for memory deficit, disturbances in concentration, depression, dizziness, vertigo and headache. Ginkgo leaf contains kaempterol-3-rhamnoglucoside, ginkgetin, isoginketin and bilobetin. Several flavone glycosides have also been isolated (ginkgolide A and B). Other substances isolated include shikimic acid, D-glucarica acid and anacardic acid. Extract of ginkgo leaf stimulates vasodilatation, resulting in a decrease in blood pressure. Several chemicals found in ginkgo extracts, especially ginkgolide B, are potent antagonists against platelet activity factor and also have an antioxidant effect. Wettstein reported that the second-generation cholinesterase inhibitors (donepezil, rivastigmine, metrifonate) and ginkgo extract are equally effective in the treatment of mild to moderate Alzheimer's dementia (8). The cognitive effects of ginkgo have also been evaluated in non-Alzheimer patients. In 18 non-demented elderly who had a slight age-related memory loss based on immediate recall of three lists of words, ginkgo extract improved the speed of information processing based on an ability to recall presented pictures and words (9). Vertigo and tinnitus have been successfully treated with ginkgo at doses of 16–160 mg/day for 3 months (10).

The most common adverse effects of ginkgo are gastric disturbances, headache and dizziness. Miwa et al. reported a case of a 36-year-old woman who had a generalized convulsion 4 h after ingestion of 70–80 gingko nuts to improve her health (11). One common adverse effect of ginkgo reported is bleeding. Spontaneous intracerebral hemorrhage occurred in a 72-year-old woman who was taking 50 mg of ginkgo three times a day for 6 months (12). Fessenden et al. reported a case of postoperative bleeding after laparoscopic cholecystectomy (13). One report described a 70-year-old man who presented with bleeding from the iris into the anterior chamber of the eye 1 week after beginning a self-prescribed regimen consisting of a concentrated ginkgo biloba extract at a dose of 40 mg twice a day. His medical history included coronary artery bypass surgery performed 3 years previously. His only medication was 325 mg of aspirin daily. After the spontaneous bleeding episode, he continued taking aspirin but discontinued ginkgo use. Over a 3-month follow-up period, he had no further bleeding episodes. Interaction between ginkgo and aspirin was considered the cause of his ocular hemorrhage (14). Hauser et al. reported a case of bleeding complications after liver transplant in a 59-year-old patient due to the use of ginkgo biloba. Seven days after a second liver transplant in this patient, subphrenic hematoma occurred. Three weeks later an episode of vitreous hemorrhage was documented. No further bleeding occurred after the patient stopped taking ginkgo biloba (15).

7.4 Garlic and ginger: efficacy and toxicity

As discussed on page 8 of this book, garlic is promoted for lowering risk of cardiovascular disease by reducing cholesterol and some clinical studies have demonstrated efficacy of garlic (16). Garlic is cheap and effective but may produce significant drug interaction with warfarin (see chapter 6). Mixture of chopped garlic on oil left in room temperature may be dangerous as discussed on page 8 (17). Ginger like garlic is also effective in preventing nausea and vomiting (18). Please see page 9 for detailed discussion of the subject.

7.5 Drug interactions with ginseng

Panax ginseng has been shown to increase the elimination of alcohol in humans, rat, and mice. Alcohol clearance increased by 30% in a non-randomized clinical study of 14 healthy volunteers when co-administered with ginseng extract compared to controls (19). Ginseng also increased the clearance of alcohol in rats when administered orally (20). However, when alcohol was given intraperitoneally, no changes were observed. In mice, administration of ginseng 10 min before ethanol consumption resulted in lower plasma ethanol concentrations, particularly in the first 30 min (21).

Two case reports describe symptoms including insomnia, headache, irritability and visual hallucinations when phenelzine was taken concurrently with ginseng (22,23). In another case, a decrease in international normalization ratio (INR) from 3.1 to 1.5 (target 2.5–3.5) was observed for a patient on warfarin 2 weeks after starting to take ginseng three times daily to increase his energy level. Two weeks after he discontinued ginseng use, his INR recovered to 3.3 (24).

There is a case report of clinically significant interference of Siberian ginseng in serum digoxin measurement by an immunoassay. However, in the experience of this author, Siberian ginseng modestly elevates serum digoxin values only if measured by a poly-clonal immunoassay and more specific monoclonal immunoassays are free from such interference. Chapter 4 contains a more detailed discussion.

7.6 Drug interactions with ginkgo biloba

Ginkgo biloba is an inhibitor of platelet-activating factor and therefore can lead to clini-cally significant interactions if used with drugs with an anticoagulant effect. There are several case reports of hemorrhage in users taking ginkgo with either warfarin or aspirin. A 65-year-old man taking both diclofenac sodium and a brown liquid herbal medicine developed an acute wound hemorrhage immediately after an elective total hip arthro-plasty, although he had been advised to discontinue use of both diclofenac and the herbal supplement for 2 weeks prior to surgery. The herbalist who had sold the herbal product confirmed that the active ingredients were ginkgo biloba, piper lango, asthma weed, and Iceland moss. The bleeding stopped after 8 h and the patient was discharged on the fifth post-operative day (25). In a second case, a 70-year-old male who had been taking aspirin (325 mg) developed spontaneous hyphema within 1 week of twice-daily ingestion of a ginkgo biloba tablet containing 40 mg of concentrated extract. On his physician's advice, the patient continued taking aspirin but stopped taking the ginkgo extract. There were no further bleeding episodes over a 3-month follow-up period (14). Ginkgo biloba can also potentiate the action of warfarin. Chapter 6 contains a more detailed discussion.

Trazodone is a second-generation antidepressant that inhibits serotonin uptake. There is a single case report of an 80-year-old woman with Alzheimer's disease who fell into a coma while taking twice-daily trazodone (20 mg) in conjunction with ginkgo biloba twice daily (26). Coma reversal was accomplished by administration of flumazenil, a specific antagonist of the benzodiazepine receptor. It is known that ginkgo biloba can interact with the benzodiazepine receptor, but trazodone has no effect on this receptor.

Omeprazole is a proton pump inhibitor used in the treatment of peptic ulcer disease. A clinical study involving 18 healthy Chinese subjects showed that twice-daily ginkgo biloba leaf extract (70 mg) significantly decreased plasma concentrations of omeprazole and its sulfone conjugate, while causing an increase in its hydroxylated form. The authors concluded that the mechanism is related to induction of CYP2C19 resulting in increased hydroxylation of omeprazole (27). Ginkgo has also been implicated in interactions with anti-hypertensives. An elderly patient on thiazide therapy for hypertension developed an increase in blood pressure while taking ginkgo. His blood pressure returned to normal when ginkgo use was discontinued (28).

7.7 Drug interactions with garlic

Cytochrome P450 mixed function oxidase is the most important phase I enzyme system responsible for the metabolism of a variety of drugs. Many HIV-infected patients take herbal medicines to supplement their conventional medical care and most antiretrovirals used in treating AIDS patients are primarily cleared by intestinal and hepatic CYP3A4. Fitzsimmons and Collins studied the interaction between a garlic supplement and the HIV protease inhibitor saquinavir in 10 healthy volunteers (4 men and 6 women) over a 3-week period. Administration of garlic caplets containing 3.6 mg of garlic powdered extract, twice daily, decreased the plasma AUC of saquinavir by 51%, the trough plasma concentration at 8 h by 49% and C_{max} by 54% (29). However, in another study no interaction between a garlic supplement and the HIV protease inhibitor ritonavir was observed (30).

Garlic exhibits complex cardiovascular effects such as antiplatelet activity and hence could potentially interact with anticoagulant and antiplatelet drugs. The combination of warfarin and garlic extract caused an increase in clotting time and INR in two separate cases. These INR increases were attributed to garlic intake since no other medication changes had occurred. In both cases clotting times were approximately doubled (31,32). The additive effect of garlic supplementation is believed to be due to unidentified garlic components that have an anticoagulant effect and thus enhance warfarin efficacy. Certain organosulfur components of garlic have been shown to inhibit human platelet aggregation in vitro and in vivo.

7.8 Drug interactions with ginger

Ginger is a potent inhibitor of thromboxane synthetase, which causes prolonged bleeding times and may also decrease platelet aggregation. Concomitant use of ginger and anticoagulants may increase the risk of bleeding in patients. Owing to the presumed effects of ginger on platelet aggregation, there is a concern that combined use of ginger and warfarin may lead to increased anticoagulation. This concern was revealed in a case report of a 75-year-old woman on chronic warfarin therapy who experienced an increase in INR complicated by epistaxis after taking a ginger supplement. After she stopped taking the ginger supplement, her INR returned to normal (33).

In other clinical studies, a synergistic effect of ginger and nifedipine on antiplatelet aggregation has been observed in normal human volunteers and hypertensive patients. It was reported that either aspirin or ginger could potentiate the antiplatelet aggregation effect of nifedipine in patients (34).

7.9 Conclusions

As mentioned earlier, ginseng (mostly Asian ginseng), ginkgo biloba, garlic and ginger supplements are widely used throughout the world. Several types of interaction are possible, and patients taking medication should be made aware of these.

References

1. Hasan, S.S., Ahmed, S.I., Bukhari, N.I., Loon, W.C. (2009) Use of complementary and alternative medicine among patients with chronic disease at outpatient clinic. Compl. Ther. Clin. Pract. 15:152–157.
2. Heller, J., Gabbay, J.S., Ghadjar, K., Jourabchi, M., O'Hara, C., Heller, M. et al. (2006) Top-10 list of herbal and supplemental medicines used by cosmetic patients: what the plastic surgeon needs to know. Plast. Reconstr. Surg. 117:436–445.
3. Chong, S.K.F. (1988) Ginseng: is there a use in clinical medicine? Postgrad. Med. 64:841–846.
4. Hikino, H., Takahasi, M., Otake, K., Konno, C. (1986) Isolation and hypoglycemic activity of eleutherans A, B, C, D, E, F and G: glycans of Eleutherococcus senticosus roots. J. Nat. Prod. 49:293–297.
5. Siegal, R. (1979) Ginseng abuse syndrome: problems with panacea. J. Am. Med. Assoc. 241:1614–1615.
6. Dega, H., Laporte, J., Frances, C., Herson, S., Choside, O. (1996) Ginseng as a cause for Steven-Johnson syndrome? Lancet 347:1344.
7. Greenspan, E.M. (1983) Ginseng and vaginal bleeding. J. Am. Med. Assoc. 249:2018.
8. Wettstein, A. (2000) Cholinesterase inhibitors and ginkgo extracts – are they comparable in treatment of dementia? Comparison of published placebo controlled efficacy of at least six months duration. Phytomedicine 6:393–401.
9. Allain, H., Raoul, P., Lieury, A., LeCoz, F., Gandon, J. (1993) Effect of two doses of ginkgo biloba extract (EGb 761) on dual-coding test in elderly subjects. Clin. Ther. 15:549–557.
10. Kleijnen, J., Knipschild, P. (1992) Ginkgo biloba. Lancet 340:1136–1139.
11. Miwa, H., Iijima, M., Tanaka, S., Mizuno, Y. (2001) Generalized convulsion after consuming a large amount of gingko nuts. Epilepsia 42:280–281.
12. Gilbert, G.J. (1997) Ginkgo biolba. Neurology 48:1137.
13. Fessenden, J.M., Wittenborn, W., Clarke, L. (2001) Gingko biloba: a case report of herbal medicine and bleeding postoperatively from a laparoscopic cholecystectomy. Am. Surg. 67:33–35.
14. Rosenblatt, M., Mindel, J. (1997) Spontaneous hyphema associated with ingestion of Ginkgo biloba extract. N. Engl. J. Med. 336:1108.
15. Hauser, D., Gayowski, T., Singh, N. (2002) Bleeding complications precipitated by unrecognized Ginkgo biloba use after liver transplantation. Transpl. Int. 15:377–379.
16. O'Hara, M., Kiefer, D., Farrell, K., Kemper, K. (1998) A review of 12 commonly used medicinal herbs. Arch. Fam. Med. 7:523–536.
17. Morse, D.L., Pickard, L.K., Guzewich, J.J., Devine, B.D., Sayegani, M. (1990) Garlic-in-oil associated botulism: episode leads to product modification. Am. J. Public Health 80:1272–1373.
18. Phillips, S., Ruggier, R., Hutchinson, S.E. (1993) Zingiber officinale (ginger) – an antiemetic for day care surgery. Anesthesia 48:715–717.
19. Lee, F.C., Ko, J.H., Park, J.K., Lee, J.S. (1987) Effects of Panax ginseng on blood alcohol clearance in man. Clin. Exp. Pharmacol. Physiol. 14:543–546.
20. Lee, Y.J., Pantuck, C.B., Pantuck, E.J. (1993) Effect of ginseng on plasma levels of ethanol in the rat. Planta Med. 59:17–19.

21. Petkov, V., Koushev, V., Panova, Y. (1977) Accelerated ethanol elimination under the effect of ginseng (experiments on rats). Acta Physiol. Pharmacol. Bulg. 3:46–50.
22. Shader, R.I., Greenblatt, D.J. (1985) Phenelzine and the dream machine: rambling and reflections. J. Clin. Psychopharmacol. 5:65.
23. Jones, B.D., Runikis, A.M. (1987) Interaction of ginseng with phenelzine. J. Clin. Psychopharmacol. 7:201–202.
24. Janetzky, K., Morreale, A.P. (1997) Probable interaction between ginseng and warfarin. Am. J. Health Syst. Pharm. 54:692–693.
25. Jayasekera, N., Moghal, A., Kashif, F., Karalliedde, L. (2005) Herbal medicines and postoperative haemorrhage. Anaesthesia 60:725–726.
26. Galluzzi, S., Zanetti, O., Binetti, G., Trabucchi, M., Frisoni, G.B. (2000) Coma in a patient with Alzheimer's disease taking low-dose trazodone and gingko biloba. J. Neurol. Neurosurg. Psychiatry 68:679–680.
27. Yin, O.Q., Tomlinson, B., Waye, M.M., Chow, A.H., Chow, M.S. (2004) Pharmacogenetics and herb-drug interactions: experience with Ginkgo biloba and omeprazole. Pharmacogenetics 14:841–850.
28. Shaw, D., Leon, C., Kolev, S., Murray, V. (1997) Traditional remedies and food supplements. A 5-year toxicological study (1991–1995). Drug Saf. 17:342–356.
29. Fitzsimmons, M.E., Collins, J.M. (1997) Selective biotransformation of the human immunodeficiency virus protease inhibitor saquinavir by human small-intestinal cytochrome P4503A4: potential contribution to high first-pass metabolism. Drug Metab. Dispos. 25:256–266.
30. Gallicano, K., Foster, B., Choudhri, S. (2003) Effect of short-term administration of garlic supplements on single-dose ritonavir pharmacokinetics in healthy volunteers. Br. J. Clin. Pharmacol. 55:199–202.
31. Sunter, W.H. (1991) Warfarin and garlic. Pharm. J. 246:772.
32. Evans, V. (2000) Herbs and the brain: friend or foe? The effects of ginkgo and garlic on warfarin use. J. Neurosci. Nurs. 32:229–232.
33. Kruth, P., Brosi, E., Fux, R., Morike, K., Gleiter, C.H. (2004) Ginger-associated overanticoagulation by phenprocoumon. Ann. Pharmacother. 38:257–260.
34. Young, H.Y., Liao, J.C., Chang, Y.S., Luo, Y.L., Lu, M.C., Peng, W.H. (2006) Synergistic effect of ginger and nifedipine on human platelet aggregation: a study in hypertensive patients and normal volunteers. Am. J. Chin. Med. 34:545–551.

8 Interactions between fruit juices and drugs: clinical impact

8.1 Introduction

Drinking fruit juices such as orange juice, apple juice and grapefruit juice is a healthy practice because such juices are rich in vitamin C and antioxidants. However, certain fruit juices, such as grapefruit juice, are known to interact with many Western drugs, which is of clinical significance. Americans regularly consume grapefruit juice, with 14% of the population drinking the juice at least weekly (1). Grapefruit juice is a well-known inhibitor of CYP3A4, which is present in the intestinal wall and is responsible for the metabolism of over 50% of commonly prescribed drugs, and thus increases the bioavailability of many drugs (1). The effect of grapefruit juice in increasing bioavailability of certain drugs may last as long as 24 h after consuming just one glass. Therefore, individuals taking certain drugs may experience drug toxicity on consuming grapefruit juice. Although interactions between grapefruit juice and various drugs are common, drug interactions with orange juice, Seville orange juice, cranberry juice, and pomegranate juice have also been reported.

8.2 Drug interactions with grapefruit juice

It was reported for the first time in 1991 that a single glass of grapefruit juice caused a two- to three-fold increase in the plasma concentration of felodipine, a calcium channel blocker, after oral intake of a standard dose of one 5-mg tablet. However, no interaction was noted after consumption of a similar amount of orange juice (2). Subsequent investigations demonstrated that the pharmacokinetics of approximately 40 other drugs is also affected by intake of grapefruit juice (3).

Furanocoumarins found in the grapefruit juice are probably responsible for interactions between grapefruit juice and drugs. The major furanocoumarin present in grapefruit juice is bergamottin, which inhibits intestinal cytochrome P450 mixed function oxidases. Moreover, a metabolite of bergamottin (6',7'-dihydrobergamottin) and a number of other furanocoumarin derivatives may also be involved in inhibiting CYP3A4 and CYP1B1 (4). The chemical structure of bergamottin is shown in ▶Fig. 8.1. De Castro et al. reported that bergamottin concentrations varied from 1.0 to 36.6 mmol/L in various commercial grapefruit juices. Therefore, the magnitude of interaction between grapefruit juice and a drug may vary, depending on the bergamottin content of the juice (5). Paine et al. reported that furanocoumarin-free grapefruit juice showed no interaction with felodipine, thus establishing that furanocoumarins are responsible for interactions between felodipine and grapefruit juice (6).

When cytochrome P450 enzymes are inhibited by components of grapefruit juice, the intestinal metabolism of some drugs is inhibited before they are absorbed. Therefore, a

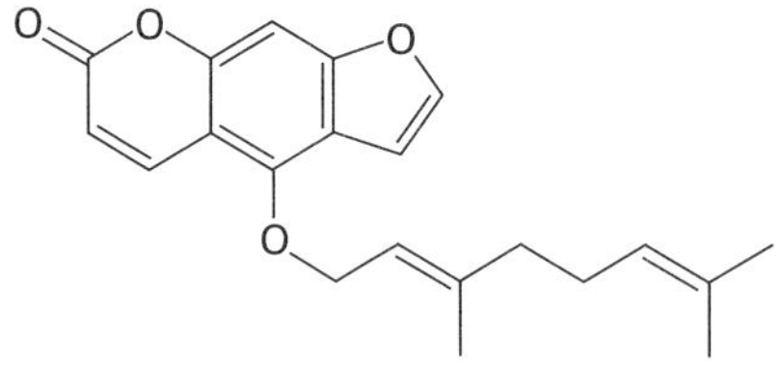

Fig. 8.1: Chemical structure of bergamottin.

greater drug amount is available for absorption from the gut. In addition to inhibiting intestinal cytochrome P450 enzymes, components of grapefruit juice also inhibit P-glycoprotein, and thus also increase the bioavailability of drugs whose absorption is modulated by P-glycoprotein, which carries drug molecules back to the intestinal lumen from enterocytes and thus decreases the fraction absorbed. Although grapefruit juice causes significant increases in the bioavailability of some drugs after oral dosing, grapefruit juice does not alter the pharmacokinetic parameters of the same drugs after intravenous administration. Therefore, it seems that grapefruit juice inhibits only intestinal and not liver cytochrome P450 enzymes.

The interaction between grapefruit juice and a drug depends on the amount of grapefruit juice consumed and the concentration of active components in the juice responsible for such interactions. In general, a standard glass of fruit juice contains 200–300 mL and ingestion of 200 mL of grapefruit juice is sufficient to cause a drug interaction. The time of juice ingestion also affects the interaction between grapefruit juice and drugs. Even if a drug is taken long after consuming grapefruit juice (6–12 h), an interaction may be still observed. However, 24 h after consuming grapefruit juice, no interaction is likely to be observed.

8.3 Drug classes that interact with grapefruit juice

Various classes of drugs, including antiallergics, antibiotics, antiarrhythmics, anticoagulants, antimalarials, antineoplastics, antiparasitics, beta-blockers, calcium channel blockers, sedative-hypnotic drugs, HIV protease inhibitors, cholesterol-lowering drugs (statins), and immunosuppressants interact with grapefruit juice. Drugs that undergo first-pass metabolism by intestinal CYP3A4, such as felodipine and amiodarone, have clinically significant interactions with grapefruit juice, which thus increases their toxicity. Grapefruit juice-drug interactions are listed in ▶Tab. 8.1.

Interactions between grapefruit juice and various calcium channel antagonists have been extensively studied. The most striking effect was observed with felodipine, for which increases up to 430% in maximum serum concentration and up to 300% in area under the curve (AUC) were observed in the presence of grapefruit juice. A significant decrease in diastolic blood pressure was also observed when felodipine was taken with grapefruit juice, as well as adverse effects such as increased heart rate and orthostatic hypotension (2). Similar interactions with grapefruit juice were observed for nisoldipine and nicardipine. Increased bioavailability of nitrendipine (100% increase), pranidipine (73% increase) and nimodipine (50% increase) was observed in the presence of grapefruit juice (7). A significant increase in verapamil bioavailability in the presence of

Tab. 8.1: Common grapefruit juice-drug interactions

Drug class	Magnitude of interaction	Individual drugs
Calcium channel blockers	Significant interaction increasing bioavailability	Felodipine, nitrendipine
	Modest interaction with increased bioavailability	Nimodipine
Cardioactive	Significant inhibition of metabolism leading to 50% increase in AUC	Amiodarone
Statins	AUC may be increased 3.6-fold	Simvastatin
	AUC increased by 83%	Atorvastatin
	Modest increase in AUC	Pitavastatin, lovastatin
Benzodiazepines	Significant increase in blood levels	Diazepam, triazolam, midazolam
Anticonvulsants	Increased blood level	Carbamazepine
Immunosuppressants	Modest to significant increase in blood levels	Cyclosporine, tacrolimus
Protease inhibitors	Significant increase in level	Saquinavir

grapefruit juice has also been reported. Pillai et al. reported a case of severe verapamil toxicity in a 42-year-old woman who ingested only three sustained-release verapamil tablets (120 mg each) and consumed grapefruit juice (1).

Grapefruit juice interacts with cholesterol lowering drugs (statins) which act by inhibiting HMG-CoA reductase (3-hydroxy-3-methylglutaryl-coenzyme A), a key step in cholesterol synthesis by the liver. Simvastatin, a substrate for CYP3A4, is extensively metabolized during first pass. Grapefruit juice (200 mL once a day for 3 days) increased the simvastatin AUC (0–24 h) 3.6 fold. Peak concentrations also increased significantly (8). In addition, when subjects ingested 200 mL of double-strength grapefruit juice three times a day for 3 days, peak serum concentrations and AUC increased 12.0- and 13.5-fold, respectively, compared to the control. When simvastatin was taken 24 h after ingestion of grapefruit juice, the effect on simvastatin AUC was only 10% of that observed when simvastatin and grapefruit juice were taken concomitantly. In addition to simvastatin, which shows the most significant interaction with grapefruit juice, other statins such as lovastatin and atorvastatin interact with grapefruit juice, but statins such as pravastatin, rosuvastatin and pitavastatin, which are minimally metabolized and are excreted mostly unchanged in urine, show little or no significant interaction with grapefruit juice. Grapefruit juice increased the atorvastatin acid AUC (0–24 h) by 83% whereas the pitavastatin AUC increased by only 13% (9). One of the major toxic effects of statins is rhabdomyolysis. Statin-associated rhabdomyolysis is triggered by grapefruit juice consumption.

Grapefruit juice interacts with the immunosuppressant drugs cyclosporine and tacrolimus. In a randomized study in 10 renal transplant recipients taking cyclosporine, Hermann et al. observed that administration of cyclosporine with grapefruit juice compared to water significantly increased the cyclosporine AUC (0–12 h) by 25% on average. However, grapefruit juice did not alter the maximum whole-blood cyclosporine concentration. The authors concluded that co-administration of cyclosporine and grapefruit juice induced a moderate but significant increase in systematic cyclosporine exposure in renal transplant recipients (10). Liu et al. studied the interaction of grapefruit juice with tacrolimus in 30 liver transplant recipients using two different types of grapefruit juice. Whole-blood tacrolimus levels were measured in the experimental groups and a control group on day 7 after starting grapefruit juice ingestion. The authors observed significant increases in tacrolimus bioavailability due to intake of grapefruit juice and concluded that tacrolimus concentrations should be closely monitored and adjusted in patients consuming grapefruit juice (11).

Protease inhibitors such as indinavir do not interact with grapefruit juice, but saquinavir, which has poor bioavailability due to extensive metabolism by CYP3A4 in the intestine, exhibits significant interactions with grapefruit juice. The bioavailability of oral saquinavir increased by a factor of two following ingestion of 400 mL of grapefruit juice without affecting clearance. This effect was absent when the drug was delivered intravenously (12).

Interactions between grapefruit juice and antiepileptic drugs have been reported. In a randomized crossover study involving 10 patients with epilepsy, 300 mL of grapefruit juice increased the trough carbamazepine concentration (4.51 μg/mL in the control group versus 6.28 μg/mL in the grapefruit juice group) (therapeutic concentration 4–12 mg/mL). Steady-state carbamazepine concentrations also significantly increased in patients who ingested grapefruit juice and carbamazepine in comparison to the control group (13). However, grapefruit juice does not interact with phenytoin.

Grapefruit juice increases the bioavailability of several benzodiazepines, including diazepam, triazolam and midazolam, but has no effect on alprazolam, even after repeated intake. Ozdemir et al. reported a threefold increase in diazepam AUC due to intake of grapefruit juice (14). Sugimoto et al. reported that grapefruit juice increased plasma concentrations of both triazolam and quazepam and of the active metabolite of quazepam, 2-oxoquazepam. The triazolam AUC significantly increased by 98%. By contrast, the quazepam AUC increased by 38%. The authors concluded that grapefruit juice has a greater effect on triazolam than on quazepam pharmacodynamics because triazolam is pre-systematically metabolized by CYP3A4, whereas quazepam is pre-systematically metabolized by both CYP3A4 and CYP2C9 (15).

Digoxin is also a substrate of P-glycoprotein but there is no significant interaction between digoxin and grapefruit juice. However, grapefruit juice significantly inhibits the metabolism of another cardioactive drug, amiodarone. Libersa et al. demonstrated that ingestion of grapefruit juice completely inhibited production of N-desethylamiodarone, the major metabolite of amiodarone, in 11 healthy volunteers and thus increased the amiodarone AUC by an average of 50% and maximum serum amiodarone by an average of 84% compared to control subjects (16).

8.4 Drug interactions with orange juice

Although orange juice does not have furanocoumarins, like grapefruit juice, several interactions between orange juice and various drugs have been reported. Orange juice increased the AUC (0–4 h) of pravastatin (a 3-hydroxy-3-methyl glutaryl CoA reductase inhibitor) in healthy volunteers when administered orally. However orange juice had no effect on the bioavailability of simvastatin, another 3-hydroxy-3-methyl glutaryl

CoA reductase inhibitor (17). Orange juice also enhances aluminum absorption from antacid preparations. In a study involving 15 normal subjects, concomitant ingestion of orange juice with the antacid "Aludrox" led to an approximately 10-fold increase in 24-h urinary excretion of aluminum (18). In contrast to these observations of increased bioavailability, other studies indicate that orange juice can reduce the bioavailability of certain drugs. Kamath et al. reported that orange juice and apple juice significantly reduced the bioavailability of oral fexofenadine in rats (by 31% and 22%, respectively). This may be related to inhibition of organic anion transporter polypeptides (19). Orange juice substantially reduced the bioavailability of celiprolol, a β-adrenergic receptor blocking agent used in the treatment of hypertension. In a study involving 10 healthy volunteers, orange juice (200 mL) reduced the mean peak plasma concentration of celiprolol by 89%. The AUC also decreased by 83% due to intake of orange juice. The authors concluded that this interaction is likely to have clinical importance (20).

Sour orange, also known as bitter or Seville orange, differs from the regular sweet orange that is most commonly consumed. In contrast to orange juice, Seville orange juice contains furonocoumarins and therefore can interact with drugs in a manner similar to that observed for grapefruit juice. Malhotra et al. reported that Seville orange juice increased the felodipine AUC by 76%, whereas grapefruit juice increased the AUC by 93%. The concentrations of bergamottin and 6,7-dihydroxybergamottin were 5 and 36 mmol/L in Seville orange juice and 16 and 23 mmol/L in grapefruit juice, respectively. The authors concluded that Seville orange juice and grapefruit juice interact with felodipine via the same mechanism (21). Drug interactions with orange juice, Seville orange juice and other fruit juices are listed in ▶Tab. 8.2.

8.5 Drug interactions with other fruit juices

Components of pomegranate juice are potent inhibitors of CYP3A4. Komperda described a case in which a 64-year-old Caucasian woman being treated with warfarin for recurrent deep vein thrombosis showed a stable international normalization ratio (INR) for several months on a daily warfarin dosage of 4 mg. At the time she was drinking pomegranate

Tab. 8.2: Interaction of various fruit juices with drugs

Fruit juice	Magnitude of drug interaction	Individual drug
Orange juice	Significant increase in AUC AUC reduced by 83%	Pravastatin Celiprolol
Seville orange juice	AUC increased by 76%	Felodipine
Pomelo juice	Increased blood level Reduced blood level	Cyclosporine, tacrolimus Sildenafil
Pomegranate juice	AUC increased 1.5-fold Increased effectiveness	Carbamazepine Warfarin
Cranberry juice	INR >50 in a patient causing death	Warfarin

juice two to three times a week. When she stopped drinking pomegranate juice, her INR decreased to a sub-therapeutic level requiring an increase in warfarin dose to maintain therapeutic anticoagulation (22).

Pomelo is a citrus fruit native to Southeast Asia and is pale green to yellow when ripe. It is the largest citrus fruit and can weigh up to 2 kg. It is also called pummelo, pommelo and Chinese grapefruit. Pomelo, which is closely related to grapefruit, interacts with cyclosporine because it contains furanocoumarins. In a study involving 12 healthy male volunteers, when 200 mg of cyclosporine was administered with 240 mL of pomelo juice, the average maximum concentration of cyclosporine in whole blood was 1494 ng/mL. By contrast, when same cyclosporine dose was administered with water, the average maximum concentration was 1311 ng/mL (Therapeutic concentration 75–300 ng/mL depending on transplant type. If multiple immunosuppressants are used, lower level may be targeted.). However, intake of cyclosporine with cranberry juice had no effect on cyclosporine bioavailability (23). Pomelo juice also increases blood levels of tacrolimus. Egashira et al. reported a case in which a 44-year-old male showed therapeutic tacrolimus concentrations 3 months after a renal transplant. His tacrolimus concentration varied between 8 and 10 ng/mL but on one occasion his blood tacrolimus concentration was 25.2 ng/mL. There had been no change in tacrolimus dose. The patient had consumed approximately 100 g of pomelo that was grown in his garden (24). Although ingestion of pomelo juice increases blood levels of both cyclosporine and tacrolimus, it reduced sildenafil citrate bioavailability in healthy male volunteers by approximately 60%. The maximum plasma sildenafil concentration decreased from an average of 212.4 to 134.1 ng/mL due to co-administration of sildenafil and pomelo juice. The authors concluded that individuals should not consume pomelo juice immediately before or after taking sildenafil (25).

Cranberry juice is a potent inhibitor of human cytochrome P450 mixed function oxidases. Interaction between cranberry juice and warfarin has also been reported. After a chest infection, a male patient experienced poor appetite and ingested nothing but cranberry juice for 2 weeks, as well as his regular drugs (digoxin, phenytoin and warfarin). Six weeks after starting to drink cranberry juice, he was admitted to hospital with an INR of >50. He died of gastrointestinal and pericardial hemorrhage. Cranberry juice contains flavonoids that can inhibit cytochrome P450 and warfarin is predominately metabolized by CYP2C19. The authors concluded that patients taking warfarin should not consume cranberry juice (26).

8.6 Role of therapeutic drug monitoring in identifying fruit juice-drug interaction

Interactions between various fruit juices and cyclosporine, tacrolimus, carbamazepine and warfarin can be identified from unexpected therapeutic drug levels (or INR for warfarin) during routine monitoring. However, the majority of drugs that interact with grapefruit juice and other fruit juices are not commonly monitored and may be more difficult to identify unless interactions cause severe clinical symptoms. Patients usually consider drinking fruit juice as a health practice and rarely discuss such consumption with their physician. Therefore, it is important to ask patients about their consumption of fruit juices because therapeutic drug monitoring is capable of identifying relatively few interactions between various fruit juices and drugs.

8.7 Conclusions

Although interactions between grapefruit juice and various drugs have been studied extensively and many such interactions are of clinical significance, interactions between other fruit juices and various drugs also have clinical importance. The major mechanism of such interactions is inhibition of CYP3A isoenzyme. Kim et al. demonstrated that the inhibitory potential for human CYP3A decreased in the order grapefruit>black mulberry>wild grape>pomegranate>black raspberry. Although drug interactions with pomegranate juice have been reported, black mulberry and black raspberry juices also have potential to interact with drugs (27). Grapefruit juice and Seville orange juice cause clinically significant drug interactions by inhibiting CYP3A, whereas grapefruit juice, orange juice and apple juice inhibit organic anion transporter polypeptides and thus interact with various drugs (28).

Because grapefruit juice and Seville orange juice cause most of the clinically important drug interactions, especially with immunosuppressants, it is advisable to educate transplant recipients regarding such interactions and encourage them not to consume grapefruit juice or even any herbal supplement owing to the possibility of drug-herb interactions. Similarly, patients attending a warfarin clinic should also be discouraged from taking any herbal supplement and consuming grapefruit and possibly other fruit juices.

References

1. Pillai, U., Muzaffar, J., Sen, S., Yancey, A. (2009) Grapefruit juice and verapamil toxicity: a toxic cocktail. South. Med. J. 102:308–309.
2. Bailey, D.G., Spence, J.D., Munoz, C., Arnold, J.M. (1991) Interaction of citrus juices with felodipine and nifedipine. Lancet 337:268–269.
3. Saito, M., Hirata-Koizumi, M., Matsumoto, M., Urano, T., Hasegawa, R. (2005) Undesirable effects of citrus juice on the pharmacokinetics of drugs: focus on recent studies. Drug Saf. 28:677–694.
4. Girennavar, B., Poulose, S.M., Jayaprakasha, G.K., Bhat, N.G., Patil, B.S. (2006) Furocoumarins from grapefruit juice and their effect on human CYP 3A4 and CYP 1B1 isoenzymes. Bioorg. Med. Chem. 14:2602–2612.
5. De Casto, W.V., Mertens-Talcott, S., Rubner, A., Butterweck, V., Dernedorf, H. (2006) Variation of flavonoids and furanocoumarins in grapefruit juice: a potential source of variability in grapefruit juice-drug interaction studies. J. Agric. Food Chem. 54:249–255.
6. Paine, M.F., Widmer, W.W., Hart, H.L., Pusek, S.N. (2006) A furanocoumarin-free grapefruit juice establishes furanocoumarins as the mediators of the grapefruit juice-felodipine interaction. Am. J. Clin. Nutr. 83:1097–1105.
7. Uno, T., Ohkubo, T., Sugawara, K., Higashiyama, A., Motomura, S., Ishizaki, T. (2000) Effects of grapefruit juice on the stereoselective disposition of nicardipine in humans: evidence for dominant presystematic elimination at the gut site. Eur. J. Clin. Pharmacol. 56:643–649.
8. Lilja, J.J., Neuvonen, M., Neuvonen, P.J. (2004) Effects of regular consumption of grapefruit juice on the pharmacokinetics of simvastatin. Br. J. Clin. Pharmacol. 58:56–60.
9. Ando, H., Tsuruoka, S., Yanagihara, H., Sugimoto, K., Miyata, M., Yamazoe, Y. et al. (2005) Effects of grapefruit juice on the pharmacokinetics of pitavastatin and atorvastatin. Br. J. Clin. Pharmacol. 60:494–497.
10. Hermann, M., Asberg, A., Reubsaet, J.L., Sather, S., Berg, K.J., Christensen, H. (2002) Intake of grapefruit juice alters the metabolic pattern of cyclosporine A in renal transplant recipients. Int. J. Clin. Pharmacol. Ther. 40:451–456.

11. Liu, C., Shang, Y.F., Zhang, X.F., Zhang, X.G., Wang, B., Wu, Z. et al. (2009) Co-administration of grapefruit juice increases bioavailability of tacrolimus in liver transplant patients: a prospective study. Eur. J. Clin. Pharmacol. 65:881–885.

12. Kupferschmidt, H.H., Fattinger, K.E., Ha, H.R., Follath, F., Krahenbuhl, S. (1998) Grapefruit juice enhances the bioavailability of HIV protease inhibitor saquinavir in man. Br. J. Clin. Pharmacol. 45:355–359.

13. Garg, S.K., Kumar, N., Bhargava, V.K., Prabhakar, S.K. (1998) Effect of grapefruit juice on carbamazepine bioavailability in patients with epilepsy. Clin. Pharmacol. Ther. 64:286–288.

14. Ozdemir, M., Aktan, Y., Boydag, B.S., Cingi, M.I., Musmul, A. (1998) Interaction between grapefruit juice and diazepam in humans. Eur. J. Drug Metab. Pharmacokinet. 23:55–59.

15. Sugimoto, K., Araki, N., Ohmori, M., Harada, K., Cui, Y., Tsuruoka, S. et al. (2006) Interaction between grapefruit juice and hypnotic drugs: comparison of triazolam and quazepam. Eur. J. Clin. Pharmacol. 62:209–215.

16. Libersa, C.C., Brique, S.A., Motte, K.B., Caron, J.F., Guedon-Moreau, L.M., Humbert, L. et al. (2000) Dramatic inhibition of amiodarone metabolism induced by grapefruit juice. Br. J. Clin. Pharmacol. 49:373–378.

17. Koitabashi, Y., Kumai, T., Matsumoto, N., Watanabe, M., Sekine, S., Yanagida, Y. et al. (2006) Orange juice increased the bioavailability of pravastatin, 3-hydroxy-3-methylglutaryl CoA reductase inhibitor, in rats and healthy human subjects. Life Sci. 78:2852–2859.

18. Fairweather-Tait, S., Hickson, K., McGaw, B., Reid, M. (1994) Orange juice enhances aluminum absorption from antacid preparation. Eur. J. Clin. Nutr. 48:71–73.

19. Kamath, A.V., Yao, M., Zhang, Y., Chong, S. (2005) Effect of fruit juices on the oral bioavailability of fexofenadine. J. Pharm. Sci. 94:233–239.

20. Lilja, J.J., Juntti-Patinen, L., Neuvonen, P.J. (2004) Orange juice substantially reduced the bioavailability of the beta-adrenergic blocking agent celiprolol. Clin. Pharmacol. Ther. 75: 184–190.

21. Malhotra, S., Bailey, D.G., Paine, M.F., Watkins, P.B. (2001) Seville orange juice-felodipine interaction: comparison with dilute grapefruit juice and involvement of furonocoumarins. Clin. Pharmacol. Ther. 69:14–23.

22. Komperda, K.E. (2009) Potential interaction between pomegranate juice and warfarin. Pharmacotherapy 29:1002–1006.

23. Grenier, J., Fradetta, C., Morelli, G., Merritt, G.J., Vranderick, M., Ducharme, M.P. (2006) Pomelo juice but not cranberry juice affects the pharmacokinetics of cyclosporine in humans. Clin. Pharmacol. Ther. 79:255–262.

24. Egashira, K., Fukuda, E., Onga, T., Yogi, Y., Matsuya, F., Koyabu, N. et al. (2003) Pomelo-induced increase in the blood level of tacrolimus in a renal transplant patient. Transplantation 75:1057.

25. Al-Ghazawi, M.A., Tutunji, M.S., AbuRuz, S.M. (2010) The effects of pummelo juice on pharmacokinetics of sildenafil in healthy adult male Jordanian volunteers. Eur. J. Clin. Pharmacol. 66:159–163.

26. Suvarna, R., Pirmohamed, M., Henderson, L. (2003) Possible interaction between warfarin and cranberry juice. Br. Med. J. 327:1454.

27. Kim, H., Yoon, Y.J., Shon, J.H., Cha, I.J., Shin, J.G., Liu, K.H. (2006) Inhibitory effects of fruit juices on CYP3A activity. Drug Metab. Dispos. 34:521–523.

28. Farkas, D., Greenblatt, D.J. (2008) Influence of fruit juices on drug disposition: discrepancies between in vitro and clinical studies. Expert Opin. Drug Metab. Toxicol. 4:381–393.

9 Heavy metal toxicity due to use of Oriental and Ayurvedic medicines

9.1 Introduction

Throughout history, herbal remedies were the only medicines available to treat various illnesses and even today the general population widely uses herbal supplements. Although many herbal supplements are safe and the toxicity of certain herbal supplements is well documented, heavy metal contamination of otherwise safe supplements poses a significant health risk. Most commonly, herbal supplements from certain Asian countries (traditional Chinese medicines) and Indian Ayurvedic medicines are contaminated with heavy metals. Lead poisoning is most the commonly reported problem after consumption of contaminated herbal products, although a significant number of case reports involve arsenic, mercury or cadmium poisoning.

9.2 Heavy metal toxicity in man

Lead poisoning is the most common type of heavy metal poisoning reported after consumption of traditional Asian medicines. Lead has no known physiological role in the human body but has a unique ability to mimic other physiologically important metal ions such as iron, zinc and calcium. Because of this ability, lead can bind to important enzymes and proteins and thus disrupts their normal biological functions. Lead interferes with biosynthesis of heme, which is essential for production of hemoglobin, and thus causes moderate to severe anemia depending on the blood burden of lead. Lead poisoning sometimes mimics the symptoms and biochemical parameters observed for porphyria, but lead poisoning is also associated with anemia, which is not usually observed in patients with porphyria. The human body eliminates lead very slowly and thus lead tends to deposit in bones. Symptoms of lead poisoning include abdominal pain, nausea, vomiting, lethargy, a metallic taste in the mouth, weight loss and neurological problems. Chronic lead poisoning in children is more dangerous than in adults because lead interferes with normal development and can negatively affect intelligence. Lead poisoning is treated by appropriate chelation therapy. It is also important to identify the source of lead exposure and remove the source from the patient's environment. In the case of lead toxicity due to ingestion of traditional Asian medicines, discontinuation of the medication is essential to eliminate the source of lead exposure, but some patients experiencing a heavy body burden of lead after consumption of a contaminated herbal supplement may require chelation therapy.

Arsenic poisoning has a lethal effect by inhibiting essential enzymes responsible for proper physiological functions. Symptoms include violent stomach pain, excessive saliva production, hoarseness and difficulty in speech. Severe poisoning leads to delirium and death. Therefore, chronic arsenic poisoning may be fatal if untreated. There are excellent chelating agents available to treat arsenic poisoning. One source of arsenic poisoning is drinking of well water if the groundwater is contaminated with arsenic. The largest mass poisoning due to arsenic in drinking water occurred in Bangladesh. In the 1970s and 1980s, international development organizations including UNICEF sponsored installation of more than 10 million wells in Bangladesh so that villagers did not need to drink water from rivers, ponds or surface water. Unfortunately, soils were not tested and it was later discovered that well water was contaminated with arsenic. In some water samples the arsenic level was five or ten times more than the safety limit recommended by the World Health Organization (WHO). A 1999 survey revealed that 27 million people were exposed to high levels of arsenic and another 50 million people were exposed to moderate amounts of arsenic over many years. The symptoms of chronic exposure to arsenic start with blackening of the hands and feet, which then spreads over the body, causing sores and even gangrene. Eventually arsenic leads to cardiovascular and reproductive damage, bladder cancer, lung cancer and other types of cancer causing serious illness and death. Arsenic exposure in children causes learning disability (1). Arsenic poisoning from drinking contaminated water has also been reported in India, China, Taiwan, Thailand, Chile and Argentina, and drinking of contaminated water is a serious public health risk in certain developing countries (2).

Mercury poisoning may occur on exposure to elemental, inorganic or organic mercury. Other than contaminated Asian medicines, a major source of mercury exposure is ingestion of contaminated fish. Fish living in contaminated lakes often build up organic mercury such as methyl mercury in their bodies from contaminated food sources. Common symptoms of mercury poisoning are peripheral neuropathy (severe itching, burning or pain), skin discoloration (pink cheeks, toes and fingernails), and swelling of the body, among others. Mercury poisoning also requires quick medical attention.

9.3 Percentage of herbal supplements contaminated with heavy metals

In 2004, Saper et al. reported on the presence of heavy metals in Ayurvedic herbal preparations purchased within a 20-mile radius of Boston. Using X-ray fluorescence spectroscopy, the authors found significant amounts of heavy metal (lead, mercury, and arsenic) contamination in 14 of 70 (20%) herbal supplements analyzed. The authors concluded if taken as recommended by the manufacturers, each of these 14 preparations could result in heavy metal intake above the regulatory limits and could result in toxicity (3). In a later study, Saper and colleagues analyzed 193 Ayurvedic medicines purchased from 25 Internet websites for manufacturers in India and the USA. Similar to their earlier study, 20.7% of all products contained heavy metals. The prevalence of metals in US manufactured products was 21.7%, compared with 19.5% in Indian products. Of the US products, 21% contained lead, 3% contained mercury, and 3% contained arsenic. Among the Indian supplements, 17% had lead, 7% had mercury, and none had arsenic (▶Tab. 9.1) (4). Patients with lead poisoning due to intake of Ayurvedic medicines often had much higher blood lead levels than patients exposed to lead-based paints and experienced greater lead toxicity (5). The presence of detectable amounts of heavy metals in Ayurvedic medicines produced in the US may be because heavy metals contained in the active components. Ayurvedic medicines containing bhasma as part of the formulation are an example of such medicinal preparations.

Tab. 9.1: Ayurvedic medicines contaminated with heavy metals (4)

Ayurvedic product	Manufacturer	Heavy metals detected
US manufacturers		
Prana-Breath of Life	Ayurherbal Corporation	Lead, mercury
AyurRelief	Balance Ayurvedic Products	Lead
GlucoRite	Balance Ayurvedic Products	Lead
Mahasudarshan	Banyan Botanicals	Lead
Kanchanar Guggulu	Banyan Botanicals	Lead
Shilajit	Banyan Botanicals	Lead
Acnenil	Bazaar of India	Lead
Energize	Bazaar of India	Lead
Hingwastika	Bazaar of India	Mercury
Bakuchi	Bazaar of India	Lead
Brahmi	Bazaar of India	Lead
Cold Aid	Bazaar of India	Lead
Trifala Guggulu	Bazaar of India	Lead, mercury, arsenic
Hear Plus	Bazaar of India	Lead
Jatamansi	Bazaar of India	Lead
Kanta Kari	Bazaar of India	Lead
Licorice	Bazaar of India	Lead
Praval Pisti	Bazaar of India	Lead, mercury
Prostate Rejuvenation	Bazaar of India	Lead
Sugar Fight	Bazaar of India	Lead
Tagar	Bazaar of India	Lead
Yograj Guggulu	Bazaar of India	Lead, arsenic
Lean Plus	Tattva's Herbs	Lead
Neem Plus	Tattva's Herbs	Lead
Indian manufacturers		
Commiphora Mukul	Unknown	Mercury
Bacopa Monniera	Unknown	Lead
Yogaraj Guggulu	Unknown	Mercury
Ezi Slim	Goodcare Pharma	Lead
Ekangvir Ras	Baidyanath	Lead, mercury[a]
Agnitundi Bati	Baidyanath	Lead, mercury[b]

(*Continued*)

(*Continued*)

Ayurvedic product	Manufacturer	Heavy metals detected
Brahmi	Baidyanath	Lead
Amoebica	Baidyanath	Lead
Arogyavardhini Bati	Baidyanath	Lead, mercury[b]
Vital Lady	Maharishi Ayurveda	Lead
Worry Free	Maharishi Ayurveda	Lead
Ayu-Arthi-Tone	Sharangdhar Pharmaceuticals	Lead
Ayu-Hemoridi-Tone	Sharangdhar Pharmaceuticals	Lead
Ayur-Leuko-Tone	Sharangdhar Pharmaceuticals	Lead
Ayu-Nephro-Tone	Sharangdhar Pharmaceuticals	Lead

[a]Very high lead and content.
[b]Extremely high mercury content.

Bhasmas are combinations of metals, herbal juices, fruits and even gemstones. The metallic chemicals found in these products include calcium, iron, zinc, mercury, silver, potassium, arsenic, copper, tin, and trace amounts of gold. It is believed that, as used, the metals exist as nanoparticles and are rendered non-toxic when they complex with components of the medicinal herbs. The basic materials in some bhasmas are listed in ▶Tab. 9.2. Although arsenic and copper are not typically included, these formulations may nevertheless be contaminated by significant amounts of these metals. Siddha Makaradhwaja contains mercury in sulfide form as an active component (6).

In another report, the authors analyzed many herbal products used in Brazil and observed the presence of cadmium, mercury and lead (7). Cooper et al. reported that

Tab. 9.2: Ayurvedic medicines (bhasma) that contain heavy metals

Bhasma	Basic material[a]
Shanka	Sea products
Swarna	Gold
Mukta	Pearl
Abrak	Manganese
Godanti	Gypsum stones
Loha	Iron
Trivang	Aluminum and zinc
Naga	Lead
Parad	Mercury

[a]Although arsenic, copper or lead are not basic ingredients of these bhasmas, significant amounts of such metals can be found in some preparations.

out of 247 traditional Chinese medicines, 5–15% of the products were contaminated with arsenic, 5% were contaminated with lead and 65% were contaminated with mercury. Some preparations even exceeded the tolerable daily intake for males and females for arsenic (4 products for males, 5 products for female), lead (1 product for male and 2 for females) and mercury (5 products for male and 7 products for females). These excesses were as high as 2760-fold, and thus pose a substantial risk of poisoning from consumption of these herbal supplements. (8). However, only a small fraction of the traditional Chinese medicines tested were contaminated with high amounts of heavy metals posing a serious health risk. Obi et al. reported that 100% of all Nigerian herbal supplements analyzed were contaminated with heavy metals, including cadmium, copper, iron, nickel, selenium, zinc, lead and mercury. The authors warned of the potential heavy metal toxicity for use of such herbal supplements (9). In another study, the authors analyzed traditional herbal remedies purchased in the USA, Vietnam and China and observed that the Asian remedies evaluated contained arsenic, lead and mercury that ranged from toxic (49% of products) to in excess of the maximum tolerable limits (74% of products) (10).

9.4 Case studies

Case report 1: A 45-year-old Korean man developed abdominal colic, muscle pain and fatigue and was hospitalized. His blood lead was elevated to 76 µg/dL (normal level <10 µg/dL), a very toxic blood level of lead. No occupational source of lead was detected. The patient had been taking the Chinese herbal preparation Hai Ge Fen (clamshell powder) for 4 weeks prior to hospital admission. Analysis of the clamshell powder revealed the presence of lead and it was concluded that the lead poisoning was due to ingestion of Chinese medicine (11).

    Case report 2: A 37-year-old man was admitted to hospital with symptoms of weakness, dizziness, nausea, and muscle pain that had developed over the previous few weeks. Laboratory studies showed severe anemia with low hemoglobin, but a thorough medical investigation did not identify the cause. The patient then reported that he had traveled to India and visited an Ayurvedic practitioner who gave him several herbal remedies. He had also had a bowel wash with oil. On his return to the USA, he continued taking most of the herbal remedies until a few days prior to his admission to hospital. Lead poisoning was suspected and his blood lead concentration was 38 µg/dL. Analysis of the herbal supplement revealed a high lead content. The patient fully recovered after receiving chelation therapy with D-penicillamine (12).

    Case report 3: A 35-year-old woman frequently visited the emergency department with severe colic pain, vomiting and weight loss. Laboratory tests showed a low hemoglobin level indicative of severe anemia. Examination of her blood cells (basophilic stripping in red blood cell smear) suggested lead poisoning and her blood lead was elevated to a dangerous level of 140 µg/dL. After ruling out environmental exposure to lead, further investigation revealed that the woman was taking various Ayurvedic medicines. Analysis revealed that one remedy had a high lead concentration and it was confirmed that her lead poisoning was due to ingestion of Ayurvedic medicine. The patient responded to chelation therapy for treatment of lead poisoning and eventually recovered (13).

Case report 4: A 24-year-old pregnant woman who emigrated from India to Australia had low hemoglobin at 24 weeks of pregnancy. The physician suspected lead poisoning based on an analysis of her peripheral smear and analysis revealed a blood lead level of 107 µg/dL, which is in the toxic range. Chelation therapy was initiated to treat her lead poisoning, and 36 h later she gave birth to a female baby. The baby was critically ill and was unresponsive. The lead concentration in the cord blood was extremely high (157 µg/dL), confirming severe lead poisoning in the neonate. Chelation therapy was initiated immediately and the infant suffered for a long time due to maternal exposure to lead. The blood level decreased to 37.2 µg/dL by week 15 and to 19.5 µg/dL after 5 months. The source of lead was confirmed as two Ayurvedic medicines the woman took periodically over a period of 9 years following the recommendation of an Ayurvedic practitioner in India. This case illustrates how lead poisoning in a mother can critically affect the health and wellbeing of her newborn (14).

Case report 5: An 11-year-old girl showed symptoms associated with arsenic toxicity and developed portal hypertension 6 and 18 months after taking Ayurvedic medicines prescribed for epilepsy. Analysis of eight Ayurvedic medicines taken by the patient showed the presence of arsenic and her serum arsenic was elevated to 202 µg/L. After discontinuation of the Ayurvedic medicines, her symptoms resolved (15).

9.5 How herbal supplements become contaminated with heavy metals

There are two major routes by which heavy metals can be incorporated into herbal supplements: unintended inclusion during plant growth or processing of the source material, and intentional addition during preparation of the final product because heavy metals are part of the ancient formula. Most plants are capable of absorbing and accumulating toxic metals present in the soil or water in which they are grown. A wide variety of endemic and anthropogenic sources are responsible for the presence of metals in the environment, which are then incorporated into the growing plant. For example, arsenic is common in groundwater and soil often contains metals used in local industries or released in waste. Similarly, the use of metal-containing pesticides, fertilizers, and other agricultural chemicals presents further potential for the introduction of potentially toxic metals into medicinal plants.

Accumulation of heavy metals and other toxic metals such as pesticides in medicinal plants is a complex process. Within a single plant, heavy metals may be unevenly distributed among the roots, leaves and other parts. In addition, the same medicinal plant grown in different geographic locations in the same country or different countries may vary widely in the amounts of heavy metals that accumulate in the plant because a plant grown in a nonindustrial area may be completely free from heavy metals. In one study, the authors collected medicinal plants from different sites and observed that of 26 specimens analyzed, four contained lead and 17 contained cadmium above the corresponding WHO acceptable limit (16). Barthwal et al, observed that the level of heavy metal was higher in soil than in plant parts they analyzed. In addition, accumulation of heavy metals varied from plant to plant grown in soil with similar heavy metal content. The level of heavy metals differed in the same plant species collected from environmentally different sites in the same city (industrial versus residential areas) (17). In another study of Indian medicinal plants, the authors demonstrated that lead and cadmium concentrations were higher in aerial parts of the plant than in the roots. Lead concentrations in 54.3% of the samples exceeded the WHO permitted limit and the cadmium content in 77.1% exceeded the limit. The authors concluded that prolonged consumption of such medicinal plants could be detrimental to human health (18). When heavy metals are present in the soil and contaminate plants from which herbal supplements are derived, plant metabolism may be altered and thus production of secondary metabolites that often have medicinal value may be suppressed (19).

As mentioned earlier, purified metals are often added to certain herbal preparations. For example, Ayurvedic medicines prepared according to the Rasa Shastra guidelines contain higher amounts of metals compared to those prepared according to other practices. Traditional medical practitioners claim that the metal ingredients are "detoxified" prior to inclusion in the remedy, but this point has not been proved by scientific research. Elevated heavy metal concentrations can obviously lead to chronic and acute illness and heavy metal poisoning requiring aggressive chelation therapy.

9.6 Pesticide contamination of herbal supplements

Although the focus of this chapter is on serious health hazards associated with ingestion of herbal supplements contaminated with heavy metals, supplement contamination with pesticides and misidentification of herbal remedies causing serious toxicity are also of significant public health interest. Contamination of vegetables and food products with pesticides is a common problem worldwide. Herbal supplements are not free from such contamination. In ancient times, pesticides were not used in cultivation of medicinal plants in China, but modern use of pesticides is inevitable in many countries owing to the increasing demand for larger quantities of herbal products worldwide. There are several reports of contamination of traditional Chinese medicines with pesticides in the literature. In one study, the authors analyzed 280 specimens of 30 different traditional Chinese medicines and observed that contamination with pesticides was widespread. However, pesticide concentrations exceeded tolerable limits for only four samples (20). The banned pesticide dichlorodiphenyltrichloroethane (DDT) was even detected in Chinese medicine (21).

A report of pesticide poisoning due to ingestion of a Chinese medicine has been documented. A 68-year-old man presented to hospital with classical symptoms of pesticide poisoning (cholinergic syndrome). Acetylcholinesterase activity in his blood cells was only 50% of the normal value, indicating pesticide poisoning. The patient was taking *Flemingia macrophylla* and ginseng (see color plate). The patient recovered after treatment with the antidote atropine. It was believed that pesticide poisoning was due to ingestion of a contaminated herb (22).

9.7 Toxicity associated with misidentified herbal supplements

Misidentification of a herbal supplement may have serious adverse effects that differ from those of a known, relatively safe herbal product. Dried herbs and roots are usually cheaper to buy than processed liquid extracts or capsules and many individuals tend to buy unprocessed herbs to save money. Unfortunately, chopped herbs are more difficult to identify and misidentification may occur. Many cases of misidentification of herbs have been reported. Plantain is an edible fruit and young plantain leaves have long been used in folk medicine to treating colds and coughs, to lower cholesterol and for other purposes. Plantain is a safe herbal product but in 1998 two case reports in the literature described serious toxicity due to the use of plantain products. Both patients had significant amounts of digoxin in their blood although neither had been treated with digoxin. Both patients experienced serious cardiac toxicity requiring hospitalization. Later it was found that 272 kg (600 lb) of plantain imported from Germany into the USA was contaminated with cardiac glycosides because the product contained *Digitalis lantana*. Digoxin is a component of *Digitalis* leaves and digoxin, along with other cardiac glycosides, caused serious toxicity effects in the two patients who ingested a nontoxic plantain product (23). Several batches of commercially available plantain were reanalyzed in 2006 and no cardiac glycoside contamination was detected (24).

Use of ginseng does not generally cause androgenization. However, maternal ingestion of Siberian ginseng caused androgenization in a newborn, who was born with thick black hair over the entire forehead and pubic area and had swollen red nipples. It was discovered that the Siberian ginseng was contaminated with Chinese silk vine (25). A 19-year-old farm labor presented to hospital with a 24-h history of abdominal discomfort and vomiting. He also experienced dizziness, blurred vision and malaise. He mistakenly ingested seven foxglove leaves thinking it was comfrey because foxglove and comfrey leaves have a similar appearance. He was not exposed to digoxin but his serum digoxin level was 3.7 ng/mL, which was consistent with foxglove poisoning; such a high digoxin level in blood is potentially fatal. Fortunately, he recovered with supportive treatment and was eventually discharged from hospital (26). This case report indicates the danger in identifying herbal plants for consumption.

In 1993, rapidly progressing kidney damage was reported in a group of young women who were taking pills containing Chinese herbs while attending a weight loss clinic in Belgium. It was discovered that one prescription Chinese herb had been replaced by another Chinese herb containing aristolochic acid, a known renal toxin. Although not related to this topic, self identification of wild mushrooms for consumption as a food is dangerous. Many *Amanita* species of mushrooms are very poisonous and may cause severe liver failure requiring liver transplant and even death. Madhok et al. reported three cases of severe liver toxicity due to ingestion of wild mushrooms. Fortunately, none of the subjects died from mushroom poisoning (27). Mushroom poisoning, which causes severe toxicity, often occurs in experienced collectors. The increasing popularity of natural foods has led to an increased incidence of wild mushroom poisoning among the general population. In one case report, a grandmother with over 30 years of self-taught expertise in identifying wild mushrooms prepared a mushroom dish for herself and three family members. A 3-year old child died from mushroom poisoning, while another recovered (28).

9.8 Contamination of herbal supplements: unintentional versus intentional

Contamination of herbal supplements with heavy metals, pesticides or Western drugs may be intentional or unintentional. For example, a group of investigators detected colchicine in placental blood samples in 5 of 24 women. All five of the women were taking herbal supplements, most notably ginkgo biloba (see color plate) and echinacea (see color plate), during pregnancy, but never took any colchicines-containing medication. The other 19 women who had no colchicine in placental blood did not take any herbal supplement. The investigators analyzed several brands of ginkgo and echinacea obtained from local herbal stores in Detroit and found measurable amounts of colchicine in the products. This was probably an accidental contamination because it was unlikely that a manufacturer would have contaminated a product with colchicine, a toxic substance (29,30). Colchicine is not usually found in these herbal supplements. However, certain adulterations of herbal products are certainly intentional. Deliberate contamination occurs when the natural ingredients required are expensive or in short supply or when a manufacturer wants to intensify a particular pharmacological action by adulterating a product using a Western drug. Use of pharmaceutical grade caffeine to supplement kola nuts is an example of intentional adulteration of a herbal supplement.

9.9 Role of the clinical laboratory

Heavy metal poisoning in a patient suspected of consuming a contaminated herbal supplement can be established by analysis of the patient's blood or urine and of the herbal product. Detection and quantitation of heavy metals can be readily performed using a variety of analytical techniques, such as colorimetric and fluorimetric assays, anodic stripping voltammetry, flame atomic absorption spectrophotometry (AAS), electrothermal AAS (also referred to as graphite furnace AAS), inductively coupled plasma atomic emission spectrometry (ICP-AES, also referred to as ICP optical emission spectrometry), and ICP-mass spectrometry. AAS, although an old method, has the advantage of being relatively inexpensive and easy to maintain, but only one metal can be analyzed at a time by this method. ICP-AES and ICP-MS have provided significant advances in multi-element analysis, and ICP-MS can also be used for isotopic abundance studies, which can determine the likely source of a metal in a biological specimen. However, such instruments are expensive.

When rapid or inexpensive analysis is required, certain screening tests are available to identify some metallic elements in various specimens. For example, the Reinsch test uses copper foil to detect arsenic, antimony, bismuth and mercury; selenium, tellurium, and sulfur may also cause a positive result. The assay is not specific but can be used as a screening test in the clinical laboratory to eliminate the possibility of the presence of a heavy metal in a herbal supplement (31). Colorimetric assays are also relatively inexpensive and simple to perform. Examples include the Gutzeit test used to analyze arsenic in urine and nickel detection using pyridoxal-4-phenyl-3-thiosemicarbazone (32,33).

Pesticide testing is not commonly available in clinical laboratories, but cholinesterase testing, which is commonly offered in many clinical laboratories, can be used as an indication of organophosphorus pesticide exposure from consumption of a contaminated herbal supplement. Cholinesterases are enzymes that catalyze hydrolysis of the neurotransmitter acetylcholine to choline and acetic acid, which is crucial step in the return of cholinergic neurons to their resting state after activation. There are two type of cholinesterase, acetylcholinesterase and pseudocholinesterase. Acetylcholinesterase hydrolyzes acetylcholine more quickly than pseudocholinesterase, whereas pseudocholinesterase hydrolyzes butyrylcholine more rapidly. These enzymes are found in serum, liver and pancreas and cholinesterase tests can also be used to determine succinylcholine sensitivity and as a measure of liver function. For a suspected misidentified herb, only a qualified botanist can help.

9.10 Conclusions

Contamination of herbal supplements with heavy metals and pesticides is a serious public safety concern. Unfortunately, neither crude plant products nor the final product before distribution are routinely tested for the presence of heavy metals. However, the most serious problem is the inclusion of heavy metals in formulations such as certain bhasma in some Ayurvedic products. Clinicians should be aware of such contamination so that if heavy metal poisoning due to consumption of a herbal supplement is suspected, proper tests can be ordered to identify any heavy metals. Patients taking multiple supplements, especially Chinese or Indian Ayurvedic medicines, are at higher risk of heavy metal toxicity from consuming such supplements.

References

1. Lowenberg, S. (2007) Scientists tackle water contamination in Bangladesh. Lancet 370:471–472.
2. Singh, N., Jumar, D., Sahu, A.P. (2007) Arsenic in the environment: effects on human health and possible prevention. J. Environ. Biol. 28 (Suppl. 2):359–365.
3. Saper, R.B., Kales, S.N., Paquin, J., Burns, M.J., Eisenberg, D.M., Davis, R.B. et al. (2004) Heavy metal content of Ayurvedic herbal medicine products. J. Am. Med. Assoc. 292:2868–2873.
4. Saper, R.B., Phillips, R.S., Sehgal, A., Khouri, N., Davis, R.B., Panquin, J. et al. (2008) Lead, mercury and arsenic in US and Indian manufactured Ayurvedic medicines sold via the Internet. J. Am. Med. Assoc. 300:915–923.
5. Kales, S.N., Christophi, C.A., Saper, R.B. (2007) Hematopoietic toxicity from lead containing Ayurvedic medications. Med. Sci. Monit. 13:CR295–298.
6. Kumar, A., Nair, A.G., Reddy, A.V., Garg, A.N. (2006) Bhasmas: unique Ayurvedic metallic-herbal preparations, chemical characterization. Biol. Trace Elem. Res. 109:231–254.
7. Caldas, E.D., Machado, L.L. (2004) Cadmium, mercury and lead in medicinal herbs in Brazil. Food Chem. Toxicol. 42:599–603.
8. Cooper, K., Noller, B., Connell, D., Yu, J., Sadler, R., Olszowy, H. et al. (2007) Public health risks from heavy metals and metalloids present in traditional Chinese medicines. J. Toxicol. Environ. Health A 70:1694–1699.
9. Obi, E., Akunyili, D.N., Ekpo, B., Orisakwe, O.E. (2006) Heavy metal hazards of Nigerian herbal remedies. Sci. Total Environ. 369:35–41.
10. Gravey, G.J., Hahn, G., Lee, R.V., Harbison, R.D. (2001) Heavy metal hazards of Asian traditional medicines. Int. J. Environ. Health Res. 11:63–71.
11. Markowitz, S.B., Nunez, C.M., Klitzman, S., Munshi, A.A., Kim, W.S., Eisinger, J. et al. (1994) Lead poisoning due to Hai Ge Fen. The porphyrin content of individual erythrocytes. J. Am. Med. Assoc. 271:932–934.
12. Spriewald, B.M., Rascu, A., Schaller, K.H., Angerer, J., Kalden, J.R., Harrer, T. (1999) Lead induced anemia due to traditional Indian medicine: a case report. Occup. Environ. Med. 56:282–283.

13. von Vonderen, M.G., Kinkenberg-Knol, E.C., Craanen, M.E., Touw, D.J., Meuwissen, S.G., de Smet, P.A. (2000) Severe gastrointestinal symptoms due to lead poisoning from Indian traditional medicine. Am. J. Gastroenterol. 95:1591–1592.

14. Tait, P.A., Vora, A., James, S., Fitzgerald, D.J., Pester, B.A. (2002) Severe congenital lead poisoning in a preterm infant due to a herbal remedy. Med. J. Aust. 177:193–195.

15. Khandpur, S., Malhotra, A.K., Bhatia, V., Gupta, S., Sharma, V.K., Mishra, R. et al. (2008) Chronic arsenic toxicity from Ayurvedic medicines. Int. J. Dermatol. 47:618–621.

16. Baye, H., Hymete, A. (2010) Lead and cadmium accumulation in medicinal plants collected from environmentally different sites. Bull. Environ. Contam. Toxicol. 84:197–201.

17. Bartwal, J., Nair, S., Kakkar, P. (2008) Heavy metal accumulation in medicinal plants collected from environmentally different sites. Biomed. Environ. Sci. 21:319–324.

18. Dey, S., Saxena, A., Dan, A., Swarup, D. (2009) Indian medicinal herbs: a source of lead and cadmium for humans and animals. Arch. Environ. Occup. Health 64:164–167.

19. Nasim, S.A., Dhir, B. (2010) Heavy metals alter the potency of medicinal plants. Rev. Environ. Contam. Toxicol. 203:139–149.

20. Xue, J., Hao, L., Peng, F. (2008) Residues of 18 organochlorine pesticides in 30 traditional Chinese medicines. Chemosphere 71:1051–1055.

21. Leung, K.S., Chan, K., Chan, C.L., Lu, G.H. (2005) Systematic evaluation of organochlorine pesticide residue in Chinese materia medica. Phytother. Res. 19:514–518.

22. Hsieh, M.J., Yen, Z.S., Chen, S.C., Fang, C.C. (2008) Acute cholinergic syndrome following ingestion of contaminated herbal extract. Emerg. Med. J. 25:781–782.

23. Slifman, N., Obermeyer, W.R., Musser, S.M., Correll, W.A. et al. (1998) Contamination of botanical dietary supplements by Digitalis lantana. N. Engl. J. Med. 339:806–810.

24. Dasgupta, A., Davis, B., Wells, A. (2006) Effect of plantain on therapeutic drug monitoring of digoxin and thirteen other common drugs. Ann. Clin. Biochem. 43:223–235.

25. Koren, G., Randor, S., Martin, S., Daneman, D. (1990) Maternal use of ginseng and neonatal androgenization. J. Am. Med. Assoc. 264:2866.

26. Turley, A.J., Muir, D.F. (2008) ECG for physicians: a potentially fatal case of mistaken identity. Resuscitation 76:323–324.

27. Madhok, M., Scalzo, A.J., Blume, C.M., Neuschwander-Tetri, B.A., Weber, J.A., Thompson, M.W. (2006) Amanita bisporigera ingestion: mistaken identity, dose-related toxicity and improvement despite severe toxicity. Pediatr. Emerg. Care 22:177–180.

28. O'Brien, B.L., Khuu, L. (1996) A fatal Sunday brunch: Amanita mushroom poisoning in a Gulf Coast family. Am. J. Gastroenterol. 91:581–583.

29. Petty, H.R., Fernando, M., Kindzelskii, A.L., Zarewych, B.N., Ksebati, M.B., Hryhorczuk, L.M. et al. (2001) Identification of colchicine in placental blood from patients using herbal medicines. Chem. Res. Toxicol. 14:1254–1258.

30. Cole, M.R., Fetrow, C.W. (2003) Adulteration of dietary supplements. Am. J. Health Syst. Pharm. 60:1576–1580.

31. Braithwaite, R. (2004) Metals and anions. In: Moffat, A., Osselton, M.D., Widdop, B., eds. Clarke's Analysis of Drugs and Poisons, Vol. 1, 3rd ed. London, UK: Pharmaceutical Press, pp. 259–278.

32. Crawford, G.M., Tavares, O. (1974) Simple hydrogen sulfide trap for the Gutzeit arsenic determination. Anal. Chem. 46:1149.

33. Sarma, L.S., Kumar, J.R., Reddy, K.J., Thriveni, T., Reddy, A.V. (2008) Development of highly sensitive extractive spectrophotometric determination of nickel(II) in medicinal leaves, soil, industrial effluents and standard alloy samples using pyridoxal-4-phenyl-3-thiosemicarbazone. J. Trace Elem. Med. Biol. 22:285–295.

10 Contamination of herbal supplements with Western drugs

10.1 Introduction

Herbal remedies are plants and plant extracts that have been used for medicinal purposes by Chinese, Indian, European, Middle Eastern, African and Native American cultures. Herbal remedies are used by the general population worldwide and in developing countries the majority of the population still use herbal supplements. Herbal remedies are not very well regulated and are freely available from health food stores and supermarkets and on the Internet. In addition, herbal supplements are not prepared according to the rigorous procedures used in manufacturing pharmaceuticals. Owing to poor regulations, these remedies are prone to contamination and adulteration with other substances such as metals, animal parts and conventional drugs. Remedies may also be adulterated with cheap, closely related derivatives of conventional drugs that may be still available in developing countries where these supplements are manufactured but are illegal in developed countries, where the supplements are eventually consumed. Owing to adulteration problems and a better understanding of the side effects of herbal products and food supplements, the FDA issued new rules in 2007 stipulating that Good Manufacturing Practices should be phased in for dietary supplements from 2008 to 2010. The new rules require that dietary supplements be properly labeled, free of adulterants and manufactured according to specified standards. Canada and the European Union also have rules to control dietary supplements and herbal remedies. However, herbal remedies produced in many other countries continue to suffer from contamination and adulteration problems due to poor regulations.

Chen et al. used liquid chromatography coupled with electrospray ionization tandem mass spectrometry to identify Western drugs in 35 out of 105 herbal supplements analyzed. Twenty three different drugs exhibiting various pharmacological actions, including anti-hypertensives, lipid-lowering agents, sedatives, hypoglycemic agents, weight-reducing agents and aphrodisiac compounds, were identified (1).

In one case study, a 33-year-old patient who had suffered from epilepsy for 8 years was treated with valproic acid, carbamazepine and phenobarbital. Prior to admission to the hospital the patient consumed three proprietary Chinese medicines in addition to taking her prescribed Western medicines. On admission, she was unable to follow any instruction and a day later she showed an unexpected serum phenytoin level of 48.5 µg/mL (194 µmol/L) (therapeutic concentration 10–120 µg/mL) although she had never been prescribed phenytoin. The toxic level of phenytoin explained her serious clinical condition. Two days later her clinical signs improved and eventually resolved 10 days after admission. Analysis of Chinese herbs showed the presence of phenytoin in one preparation, carbamazepine in another preparation and valproic acid in the third. Chinese proprietary medicines contaminated with Western anticonvulsants were responsible for her drug toxicity. Unfortunately, the labels on the Chinese medicines did not list any of these anticonvulsants as a component and no warning on potential toxicity was provided (2).

10.2 Contamination of herbal supplements with analgesics

Analgesics and non-steroidal anti-inflammatory drugs (NSAIDs) are the most commonly used over-the-counter drugs. There are a number of reports on adulteration of herbal remedies by analgesics. Liang et al. screened over 200 samples of herbal supplements obtained from the Beijing area that were suspected of being adulterated. Three out of 14 samples suspected of adulteration were positive for ibuprofen (3). Huang et al. analyzed 618 herbal supplements collected in Taiwan and suspected of adulteration and found that approximately one-quarter of the samples were adulterated with acetaminophen (167 samples, 27%) and/or indomethacin (152 samples, 25%). Other analgesics detected in herbal supplements as adulterants include phenylbutazone, oxyphenylbutazone, diclofenac, ibuprofen, ketoprofen, mefenamic acid, piroxicam and salicylamide (4). Drugs such as phenylbutazone and oxyphenylbutazone are extremely toxic and banned by the FDA. Nelson et al. observed aplastic anemia, a potentially life-threatening disease, in a 12-year-old boy who took a herbal supplement contaminated with phenylbutazone (5).

10.3 Contamination of herbal supplements with steroids

Steroids are compounds that have a steroid ring comprising three 6-carbon rings and one 5-carbon ring joined together. Examples of steroids include cholesterol, sex hormones and various drugs used in medicine, most commonly dexamethasone and prednisone used to treating inflammatory and autoimmune diseases. Anabolic steroids are synthetic drugs that mimic the effects of testosterone and corticosteroids and are used by athletes and body builders. As natural herbs may have little or no anabolic steroid activity, they may be adulterated with these drugs in an attempt to confer a similar effect. Using gas chromatography-mass spectrometry and liquid chromatography tandem mass spectrometry, Zou et al. detected dehydroepiandrosterone and androsterone in a traditional Chinese herbal product (6). Testosterone or its analogs have been found in many herbal products.

Inflammatory and autoimmune conditions such as arthritis, asthma, and inflammatory bowel disease are common disorders that are frequently treated using corticosteroids. Because these drugs are very effective in controlling inflammatory conditions, they are commonly used as adulterants in herbal remedies. Huang et al. tested 2609 samples for various drug adulterants and found 170 samples adulterated with corticosteroids, 91 with prednisolone, 23 with betamethasone, 20 with dexamethasone, 16 with prednisone, 11 with cortisone and nine with hydrocortisone (4). There are a number of case reports on patients with inflammatory diseases who developed Cushing syndrome after consuming various herbal products. Testing of the products revealed the presence of various corticosteroids, including dexamethasone, hydrocortisone and prednisolone (7). Prednisone was detected in a herbal medicine from Southeast Asia (8).

10.4 Contamination of herbal supplements with oral hypoglycemic agents

Diabetes affects millions of people worldwide and many diabetic patients take herbal supplements to improve their glycemic control. There are several reports of contamination of herbal supplement with oral hypoglycemic agents that are not listed on the product label as active ingredients. Oral hypoglycemic agents such as acetohexamide, chlorpropamide, glibenclamide and tolbutamide can be found as adulterants in some traditional Chinese medicines (9). Kulambil Padinjakara et al. described two case reports of patients with type 2 diabetes who were also taking herbal medicines. One patient was a 30-year-old postgraduate researcher who was prescribed gliclazide and metformin for her diabetes. She was also taking herbal medication without the knowledge of her physician. The other patient was a 60-year-old man with a 15-year history of type 2 diabetes who was taking insulin. In addition, this patient was also taking herbal remedies. Chemical analysis of the herbal medicines showed the presence of glibenclamide but the package inserts did not mention this Western hypoglycemic agent as an ingredient (10).

10.5 Contamination of herbal supplements with drugs that correct erectile dysfunction

Erectile dysfunction becomes a problem with advancing age and many patients are too embarrassed to discuss this problem with their physicians and take herbal supplements to correct such problems. Three phosphodiesterase type 5 inhibitors approved for the treatment of erectile dysfunction are sildenafil (Viagra™), tadalafil (Cialis™) and vardenafil (Levitra™). Since the successful introduction of these drugs, the number of herbal remedies purported to enhance sexual function has also increased in the market. Although it is claimed that they are natural products, they are often adulterated with sildenafil, tadalafil and vardenafil. The US FDA Forensic Chemistry Center analyzed 40 dietary supplements marketed as sexual enhancement formulas. Of these, 19 were positive for phosphodiesterase type 5 inhibitors used as correcting agents for erectile dysfunction; eight contained sildenafil, nine contained tadalafil and two contained the sildenafil analog, homosildenafil. This study showed that herbal products may be adulterated not only with prescription drugs, but also with their analogs, which makes detection challenging for laboratories. In addition, the adverse effects of such analogs are not fully known and thus they put consumers at unknown risk of drug toxicity (11). In another study, the authors analyzed 26 herbal remedies used for erectile dysfunction. One product contained sildenafil and 14 contained different drug analogs, including acetildenafil, hydroxyacetildenafil, hydroxyhomosildenafil and piperodenafil (12).

10.6 Contamination of herbal supplements with benzodiazepines and antidepressants

Psychoactive drugs act on the central nervous system and alter perception, mood, consciousness and behavior. Unfortunately, many herbal supplements marketed as antidepressants are contaminated with benzodiazepines or antidepressant drugs. Gertner et al.

analyzed herbal medicines taken by five different rheumatology patients and detected mefenamic acid and diazepam in the samples (13). In a large study of 2609 samples, diazepam was found in 20 samples and was the 11th most commonly used adulterant (4). Contamination of herbal supplements with both benzodiazepines and tricyclic anti-depressants has also been reported (14).

10.7 Contamination of herbal supplements with weight loss products

Obesity is a major chronic problem in the developed world. Despite changes in diet and exercise, many patients cannot control their obesity. Given the social impact of this condition, it is not surprising that obese patients purchase weight loss products, including herbal remedies that may be adulterated with conventional weight-reducing drugs. There are number of case reports and analytical data on adulteration of herbal products with weight-controlling drugs. Jung et al. reported on a 20-year-old healthy woman who purchased a Chinese herbal medicine called LiDa Dai Dai Hua Jiao Nang from the Internet for weight reduction. The patient developed headache, vertigo and numbness. The symptoms disappeared after she stopped taking the herbal remedy. The patient's urine and the herbal remedy tested positive for sibutramine, an appetite suppressant used orally to treat obesity that has been suspended by the European Union and is under review by the FDA owing to serious safety concerns. This drug causes increases in blood pressure and heart rate and interacts with many other drugs. Each capsule contained 27.4 mg of sibutramine, which is approximately twice the highest recommended dose (15). Some weight loss products contaminated with sibutramine are listed in ▶Tab. 10.1.

Herbal weight loss products imported from China are often adulterated with fenfluramine, phenolphthalein and sibutramine (5). Fenfluramine was prohibited in 1997 in the USA by the FDA because of serious toxicity, including heart failure.

10.8 Contamination of herbal supplements with other drugs

Miller and Stripp analyzed Chinese herbal medicines collected from Chinatown in New York and identified promethazine, clomethiazole, chlorpheniramine, diclofenac, chlordiazepoxide, hydrochlorothiazide, triamterene, diphenhydramine and sildenafil citrate as contaminants (16). Byard noted that drugs added to herbal supplements range from relatively safe analgesics to steroids and include aspirin, acetaminophen, antihistamines, theophylline, bromhexine, diazepam, chlordiazepoxide, diclofenac, glibenclamide, hydrochlorothiazide, mefenamic acid, aminopyrine, phenylbutazone, phenytoin, phenacetin, indomethacin and various steroids (17). Although most reports of contamination of herbal supplements with Western drugs are for traditional Chinese medicines originating from Asia. Savaliya et al. identified dexamethasone and diclofenac in 10 of 58 Ayurvedic herbal products analyzed. Another supplement tested positive for the presence of piroxicam and one supplement contained only dexamethasone (18). Western drugs of various classes commonly found in herbal supplements are listed in ▶Tab. 10.2.

Tab. 10.1: Weight loss products contaminated with sibutramine

2 Day Diet	Extrim Plus 24 Hours Reburn	Slim 3 in 1 Extra Slim Formula
2 Day Diet Slim Advance	Fasting Diet	Slim 3 in 1 Extra Slim Waist Formula
2× Powerful Slimming	Fatloss Slimming	Slim 3 in 1 M18 Royal Diet
3 Day Diet	GMP	Slim 3 in 1 Slim Formula
3 Days Fit	Imelda Fat Reducer	Slim Fast
3× Slimming Power	JM Fat Reducer	Slim Tech
5× Imelda Perfect Slimming	Lida DaiDaihua	Slim Waist Formula
7 Days Diet	Leili Meizitang	Slim Waistline
7 Diet	Miaozi Qian Ti Jiao Nang	Sliminate
8 Factor Diet	Maiozi Slim Capsules	Slimming Formula
21 Double Slim	Perfect Slim	Somotrim
24 Hours Diet	Natural Model	Super Sat Burner
999 Fitness Essence	Powerful Slim	Superslim
BioEmagrecim	Perfect Slim UP	Super Sliming
Body Creator	Perfect Slim ×5	Trim 2 Plus
Body Shaping	ProSlim Plus	Triple Slim
Body Slimming	Reduce Weight	Venom Hyperdrive 3.0
Cosmo Slim	Royal Slimming Formula	Waist Strength Formula
Extrim Plus	Sana Plus	Zhen de Shou

Tab. 10.2: Western drugs commonly encountered as adulterants in herbal supplements

Drug class	Individual drug
Analgesic/antipyretic	Acetyl salicylate, acetaminophen, phenylbutazone (banned by FDA), phenacetin, indomethacin, mefenamic acid, diclofenac
Antihistamine	Diphenhydramine, promethazine, chlorpheniramine
Anti-asthmatic	Theophylline
Anticonvulsant	Phenytoin
Diuretic	Hydrochlorothiazide, triamterene
Erectile dysfunction agent	Sildenafil citrate
Mucolytic agent	Bromhexine
Sedative/hypnotic	Chlordiazepoxide, diazepam, clomethiazole
Oral hypoglycemic agent	Glibenclamide
Steroid	Dexamethasone

10.9 Adulterated herbal supplements and the clinical laboratory

With the widespread use of herbal products and their adulteration with conventional drugs, users of these products frequently seek medical attention. Unfortunately, only a fraction of drugs found in adulterated herbal supplements are routinely monitored in clinical laboratories as a part of therapeutic drug monitoring. If a herbal product is contaminated with theophylline, phenytoin or another classical anticonvulsant or a cardioactive drug such as digoxin, procainamide or quinidine, routine testing by a clinical laboratory detect these in blood from patients suspected of taking a contaminated supplement. In addition, overdose with aspirin (acetyl salicylate) or acetaminophen (paracetamol) due to ingestion of a contaminated supplement can easily be detected because most clinical laboratories offer such tests as part of a routine toxicology panel. However, many Western drugs found as adulterants in various herbal supplements are not routinely monitored in clinical laboratories and immunoassays are not commercially available. Sophisticated chromatographic techniques coupled to mass spectrometry are required for unambiguous identification of such adulterants. In this event, a sample of the supplement and a blood or urine specimen from the patient should be sent to a reference laboratory capable of analyzing such drugs. Some drugs, such as phenyl-butazone, are no longer in use in developed countries but may be legally available in developing countries. Because these drugs are very toxic, their identification may be vital to confirm the agent as the major cause of toxicity in a patient. Unfortunately, such testing is not available even in major reference laboratories, so establishing the cause of serious toxicity in a patient due to consumption of a contaminated herbal supplement is often challenging for clinicians.

10.10 Conclusions

Although used for thousands of years by other cultures, herbal remedies have increased in popularity in the Western world in recent years. The reasons include the effectiveness of certain herbal remedies, their affordability and the general perception that herbal products are safe and harmless. Owing to poor regulations and a huge sales market, these products are often adulterated with conventional drugs or their analogs. The majority of herbal supplements in which undisclosed Western drugs have been identified as adulterants are traditional Chinese medicines and Indian Ayurvedic medicines. Herbal supplements manufactured in North America or developed countries are generally not contaminated with Western drugs. Because Chinese herbal medicines and Indian Ayurvedic medicines are readily available in herbal stores throughout the world, healthcare providers and consumers should recognize the potential risks posed by herbal remedies adulterated with conventional drugs.

References

1. Chen, Y., Zhao, L., Lu, F., Yu, Y., Chai, Y., Wu, Y. (2009) Determination of synthetic drugs used to adulterate botanical dietary supplements using QTRAP LC-MS/MS. Food Addit. Contam. A 26:595–603.

2. Lau, L.K.K., Lai, C.K., Chan, A.W. (2000) Phenytoin poisoning after using Chinese proprietary medicines. Hum. Exp. Toxicol. 19:385–386.

3. Liang, T., Hu, X., Liu, X. (2007) Investigation of banned additives in healthy foods for weight control. Chin. J. Food Hyg. 19:336–337.

4. Huang, W.F., Wen, K.C., Hsiao, M.L. (1997) Adulteration by synthetic therapeutic substances of traditional Chinese medicines in Taiwan. J. Clin. Pharmacol. 37:344–350.

5. Nelson, L., Shih, R., Hoffman, R. (1995) Aplastic anemia induced by an adulterated herbal medicine. J. Toxicol. Clin. Toxicol. 33:467–470.

6. Zou, P., Chen, P., Oh, S.S., Kiang, K.H., Low, M.Y. (2007) Detection of dehydroepiandrosterone and androsterone in a traditional Chinese herbal product. Food Addit. Contam. 24:1326–1333.

7. Hughes, J.R., Higgins, E.M., Pembroke, A.C. (1994) Oral dexamethasone masquerading as a Chinese herbal remedy. Br. J. Dermatol. 130:261.

8. Ku, Y.R., Tsai, M.J., Wen, K.C. (2001) Solid-phase extraction and high-performance liquid chromatographic analysis of prednisone adulterated in a foreign herbal medicine. J. Food Drug Anal. 9:150–152.

9. Ku, Y.R., Chag, L.Y., Ho, L.K., Lin, J.H. (2003) Analysis of synthetic anti-diabetic drugs in adulterated traditional Chinese medicines by high performance capillary electrophoresis. J. Pharm. Biomed. Anal. 33:329–334.

10. Kulambil Padinjakara, R.N., Ashawesh, K., Butt, S., Nair, R., Patel, V. (2009) Herbal remedy for diabetes: two case reports. Exp. Clin. Endocrinol. Diabetes 117:3–5.

11. Gratz, S.R., Flurer, C.L., Wolnik, K.A. (2004) Analysis of undeclared synthetic phosphodiesterase-5 inhibitors in dietary supplements and herbal matrices by LC-ESI-MS and LC-UV. J. Pharm. Biomed. Anal. 36:525–533.

12. Poon, W.T., Lam, Y.H., Lai, C.K., Chan, A.Y., Mak, T.W. (2007) Analogues of erectile dysfunction drugs: an under-recognized threat. Hong Kong Med. J. 13:359–363.

13. Gertner, E., Marshall, P.S., Filandrinos, D., Potek, A.S., Smith, T.M. (1995) Complications resulting from the use of Chinese herbal medications containing undeclared prescription drugs. Arthritis Rheum. 38:614–617.

14. Bogusz, M.J., al Tufail, M., Hassan, H. (2002) How natural are natural herbal remedies? A Saudi perspective. Adverse Drug React. Toxicol. Rev. 21:219–229.

15. Jung, J., Hermanns-Clausen, M., Weinmann, W. (2006) Anorectic sibutramine detected in a Chinese herbal drug for weight loss. Forensic Sci. Int. 161:221–222.

16. Miller, G.M., Stripp, R. (2007) A study of Western pharmaceuticals contained within samples of Chinese herbal/patent medicines collected from New York City's Chinatown. Leg. Med. (Tokyo) 9:258–264.

17. Byard, R.W. (2010) A review of the potentially forensic significance of traditional herbal medicines. J. Forensic Sci. 55:89–92.

18. Savaliya, A.A., Prasad, B., Raijada, D.K., Singh, S. (2009) Detection and characterization of synthetic steroidal and non-steroidal antiinflammatory drugs in Indian Ayurvedic/herbal products using LC-MS/TOF. Drug Test. Anal. 1:372–381.

11 Toxic and dangerous herbs

11.1 Introduction

Adverse drug reactions are a major cause of hospitalization and death in the USA. It is estimated that an average of 2.2 million people are hospitalized annually and approximately 100,000 die from adverse drug reactions (1). People turn to herbal supplements to avoid adverse drug reactions. Although many herbal supplements are fairly safe, certain herbal supplements are very toxic and may cause serious adverse reactions and even death. Dennehy et al. studied dietary-supplement-related adverse events reported to the California Poison Control System during April 2002 to September 2002 and found that most patients who experienced adverse effects from dietary supplements took ephedra-containing herbal products. The symptoms most frequently reported were increased heart rate, agitation, nausea and vomiting (2). Wang et al. reported a case of a 34-year-old female who died after using a herbal medicine powder containing *Strychnos nux-vomica* seeds. The authors identified strychnine and brucine in blood and gastric content samples from the woman. Death was attributed to use of a strychnine-containing herbal supplement (3). A 41-year-old Chinese man died from renal failure because he consumed a herbal preparation called Den Quing Wu Lin Wan that contained Guan Mu Tong as the main ingredient. Guan Mu Tong is prepared from the plant *Aristolochia manshuriensis*, which contains aristolochic acid, a known nephrotoxic compound. Microscopic examination of renal tissue revealed severe degeneration, necrosis and desquamation of renal tubular epithelial cells, the presence of protein casts and a widened edematous interstitium with interstitial fibrosis (4). There are many other case reports in the literature showing the dangers of using certain herbal supplements.

11.2 Herbal supplements that may cause death

Death may occur after ingestion of a very toxic herbal supplement or tea. A number of such reports deal with Chinese medicines such as Chan Su, aconite-containing herbal products, and products contaminated with aristolochic acid. However, death has also occurred after drinking of oleander tea or taking a kava supplement or other herbal products such as germander and chaparral. The Chinese weight loss product ma-huang, which contains ephedra alkaloids, is also very dangerous and may cause death. Herbal supplements that may cause fatality are listed in ▶Tab. 11.1.

11.2.1 Ephedra (ma-huang)

Ephedra is a small perennial shrub with thin stems that rarely grows over 1 ft (0.3 m). Some of the better-known species include *Ephedra sinica* and *Ephedra equisetina* (collectively called ma-huang) from China. Ephedrine is the predominant active compound found

Tab. 11.1: Herbal supplements that have caused fatalities

Herbal supplement	Toxic component	Cause of death
Ephedra-containing herbs	Ephedrine	Cardiac failure
Oleander tea	Oleandrin	Cardiac failure
Chan Su/Lu-Shen-Wan	Bufalin	Cardiac failure
Comfrey	Pyrrolizidine alkaloids	Liver failure
Kava	Kavalactones	Liver failure
Germander	Teucrin A	Liver failure
Pennyroyal oil	Pulegone	Liver failure
Thunder God Vine	Celastrol	Cardiac shock/damage
Monkshood	Aconite	Cardiac shock/failure
Oil of wintergreen	Methyl salicylate	Metabolic acidosis

in ephedra plants, although other compounds such as pseudoephedrine, norephedrine and phenylpropanolamine are also present in crude extracts. Ephedrine is a major component of ma-huang, a Chinese weight loss product. Ma-huang is often referred to herbal fen-phen and was once promoted as an alternative to the banned weight loss drug fen-fluramine. Sometimes herbal fen-phen products contain St. John's wort (see color plate) and are sold as herbal Prozac. Other promoted purposes include bodybuilding and enhancement of athletic performance. Herbal ecstasy is also an ephedrine-containing product that can induce a euphoric state.

There are many reports of significant toxicity after use of ephedra-containing products. Haller and Benowitz evaluated 140 reports of ephedra-related toxicity submitted to the FDA between June 1997 and March 1999 and concluded that 31% of the cases were definitely related to ephedra toxicity and another 31% were possibly related to ephedra. 47% of the reports involved cardiovascular problems and 18% involved problems with the central nervous system. Hypertension was the single most frequent adverse reaction, followed by palpitations, tachycardia, stroke and seizure. Ten events resulted in death and 13 events caused permanent disability. The authors concluded that use of a dietary supplement that contains ephedra is a serious health risk and public safety concern (5). In another case report, the authors described a 45-year-old woman who died of cardiovascular collapse after taking ephedra. Tissue analysis revealed non-specific degenerative alterations in myocardium (lipofuscin accumulation, basophilic degeneration, vacuolation of myocytes and myofibrillary loss) associated with myocyte apoptosis, caspase activation and extensive cleavage of myofibrillary proteins α-actin, α-actinine and cardiac troponin T (6).

Other than death, ephedra can increase blood pressure and can cause cardiac arrhythmia and stroke. Naik and Freudenberger reported two cases of cardiomyopathy associated with use of dietary supplements containing ephedra. A 19-year-old man presented to the emergency department with shortness of breath and chest pain radiating to the left arm. Left heart cauterization revealed no coronary artery disease. He was treated for heart failure but died 5 weeks later. A 21-year-old man also developed heart failure after taking a dietary supplement containing ephedra but responded to treatment and survived. Ephedrine is a potent sympathomimetic agent with direct and indirect effects on adrenergic receptors that cause increases in heart rate, blood pressure, cardiac output and vascular resistance. Therefore, ephedra-containing products can have a range of negative cardiovascular effects, including myocarditis, arrhythmia, myocardial infarction, cardiac arrest and sudden death (7). On April 12, 2004, the FDA prohibited the sale of ephedra-containing dietary weight loss supplements in the USA. Although banned in the USA, ma-huang is available in Asian countries and is also used for treating colds, flu, nasal congestion, asthma and fever. Use of ephedra-containing products is also banned by the International Olympic Committee.

11.2.2 Oleander tea

Oleanders are evergreen ornamental shrubs with various colors of flowers that belong to the dogbane family and grow in Southern parts of the USA, Australia, India, Sri Lanka, China and other parts of the world. All parts of oleander plants are toxic. Human exposure to oleander includes accidental exposure, ingestion by children, administration in food or drinks, for medicinal (as herbal supplements), and criminal poisoning. However, despite its toxicity, oleander is still used in some folk medicines. Boiling or drying the plant does not inactivate the toxins. A woman died after drinking herbal tea prepared from oleander (8). Oleander contains oleandrin which is similar in structure to the cardioactive drug digoxin and it causes cardiac toxicity. Because of this structural similarity, severe oleander poisoning can be treated with the same antidote (Digibind) that is used for digoxin poisoning. However, Digibind is expensive and there is a severe shortage of this antidote in countries such as Sri Lanka where death from oleander poisoning occurs (9). Oleander poisoning can be determined indirectly by taking advantage of the structural similarity between oleandrin and digoxin. Chapter 4 contains a detailed discussion on this topic.

11.2.3 Chan Su and related Asian medicines

The Chinese medicine Chan Su is prepared from dried white secretions from the auricular and skin glands of Chinese toads (*Bufo melanostictus* Schneider and *Bufo bufo gargarzinas* Gantor). Chan Su is also a major component of the traditional Chinese medicines Lu-Shen-Wan and Kyushin. These medicines are used in China for treatment of tonsillitis, sore throat, furuncles and palpitations because of their anesthetic and antibiotic actions. Chan Su given in small doses also stimulates myocardial contraction, has an anti-inflammatory effect and provides pain relief. The pharmacological effects of Chan Su are due to bufalin and cinobufagin, but bufalin is also very toxic. Death of a Chinese woman after ingestion of Chinese herbal tea containing Chan Su has been reported (10). Bufalin, the active component of Chan Su, is structurally similar to digoxin, so bufalin poisoning can be detected indirectly using digoxin immunoassays. This topic is discussed in depth in Chapter 4.

11.2.4 Death arising from use of hepatotoxic herbs

Severe hepatotoxicity may result from use of herbs such as kava, chaparral, comfrey, germander, pennyroyal and mistletoe. Several deaths have been reported due to use of the herbal sedative kava and the FDA has warned consumers regarding use of this supplement. Severe liver toxicity requiring liver transplant in a 60-year-old-woman was related to her use of chaparral for 1 year (11). Gow et al. reported a case of fatality due to acute liver failure in a patient who ingested a herbal preparation containing kava and passionflower (12). Fatal hepatitis in a 68-year-old woman was related to use of the herbal weight loss product Tealine, which contained a hepatotoxic germander extract (13). Chapter 2 contains an in-depth discussion on hepatotoxic herbs.

11.2.5 Thunder God Vine

Thunder God Vine (lei gong teng prepared from the plant *Triptergium wilfordii*) has been used in traditional Chinese medicine for over 2000 years for local treatment of arthritis and inflammatory tissue swelling. This supplement has also been indicated for treatment of rheumatoid arthritis. Unfortunately, Thunder God Vine can cause severe adverse reactions and is poisonous if not carefully extracted from the skinned roots. Other parts of the plant, such as the leaves, flowers and skin of the root, are very toxic to humans and may cause death if ingested. A 36-year-old man was admitted to hospital with severe diarrhea and vomiting for 3 days. Three days before his admission he had consumed a herbal supplement. The patient died 15 h after admission due to shock, hypotension (very low blood pressure) and cardiac damage. The herbal supplement the patient had taken was identified as Thunder God Vine (14). Celastrol, a quinine methide triterpene, is considered to be the active component of Thunder God Vine (15).

11.2.6 Aconite poisoning and fatality

Aconite, a herb native to China and Europe, is commonly known as monkshood. Aconite poisoning has been known for centuries but some Chinese herbal remedies and homeopathic remedies use aconite as an ingredient. In Chinese medicine, extracts from aconite leaves, flowers and roots are used to treat pain due to arthritis, gout, cancer, inflammation, migraine headache and sciatica. Because aconite is very toxic, its use can cause irregular heartbeat, heart block, heart failure and even death. Aconite is even dangerous to use in the form of a cream for topical application because it can be absorbed through the skin. Even touching this plant may cause an allergic reaction in allergy-sensitive persons.

The Chinese medicines Chuanwu and Caowu are prepared from the roots of various aconite species and are believed to have anti-inflammatory, analgesic and cardiotonic effects. These products contain highly toxic compounds such as aconitine, mesaconitine and hypaconitine. Death may occur from ventricular arrhythmia (very abnormal heart rhythm) usually within 24 h of ingestion of the herbal supplement. There is no antidote to aconite poisoning and only supportive therapy to sustain life is available. According to one report, three patients who consumed an aconite-containing herbal remedy died from heart failure (16). There are other reports of death in the medical literature due to ingestion of aconite-containing herbal remedies. Although accidental aconite poisoning is rare in North America, Pullela et al. described a case in which a 25-year-old man died suddenly following a recreational outing with his friends at which he ingested a number of wild berries and plants. One of the plants was later identified as monkshood. A high level of aconitine was found in post mortem blood samples, confirming the cause of death as poisoning after ingestion of monkshood plants (17).

In Chinese medicine, aconite roots are used after processing to reduce their toxicity. Soaking and boiling of roots or plant parts during processing would hydrolyze some aconite alkaloids to less toxic forms. However, inadequate processing or intake of more than the recommended dose is the major cause of toxicity after consumption of aconite-containing Chinese medicines. Patients with aconite poisoning usually present with a combination of neurological, cardiovascular and gastrointestinal features. Cardiovascular symptoms include chest pain, palpitations, bradycardia, sinus tachycardia, ventricular ectopics, ventricular tachycardia and ventricular fibrillation. The main causes of death from aconite poisoning include ventricular arrhythmias and asystole (18).

11.2.7 Poisoning and death from methyl salicylate

Methyl salicylate is a major component of oil of wintergreen, which is prepared by distillation of wintergreen leaves. Methyl salicylate has an analgesic effect and is used in many over-the-counter analgesic creams or gels designed only for topical use. Methyl salicylate if ingested is very poisonous. Aspirin is acetyl salicylate, which is structurally close to methyl salicylate, but after ingestion acetyl salicylate rapidly breaks down into salicylic acid, which is responsible for the analgesic effect. By contrast, methyl salicylate is very irritating to the gastric mucosa and very little methyl salicylate is metabolized to salicylic acid by the enzymes in human blood. The majority of methyl salicylate is converted into salicylic acid by the liver.

Oil of wintergreen contains 98% methyl salicylate and ingestion of 5 mL of oil of wintergreen (1 teaspoon) is equivalent to 7000 mg of aspirin, which could be fatal in a child (6 years old or younger) (19). The popular topical ointment Bengay® contains 15% methyl salicylate and Bengay Muscle Pin/Ultra Strength® contains 30% methyl salicylate. Many Chinese medicines and medicated oils such as white flower oil and red flower oil contain high amounts of methyl salicylate. Severe salicylate toxicity due to ingestion of white or red flower oil has been reported in adults, including one death due to ingestion of red flower oil (20). Parker et al. described a case in which a 58-year-old Vietnamese woman who lived alone in an apartment in San Diego died. Her brother discovered the body and called the emergency services. The paramedics confirmed that the woman was dead and the attending police officer discovered a cup containing a clear liquid with a menthol-like odor and several empty medicine bottles, including one Chinese herbal medicine bottle on a bedside table. All of the samples were submitted to the Medical Examiner's Office for investigation. High amounts of salicylic acid were found in blood and gastric content samples from the decedent. The Chinese herbal medicines were identified as Koong Hung Yick Far oil containing 67% oil of wintergreen and Po Sum oil containing 15% menthol. The evidence confirmed that the cause of death was ingestion of a Chinese medicated oil containing methyl salicylate (21).

Poisoning can also occur due to abuse of topical cream containing methyl salicylate. Morra et al. demonstrated that after topical application of analgesic cream containing methyl salicylate, salicylic acid could be detected in blood; the rate of methyl salicylate absorption increased in the first 24 h after application and steadily increased from day 1 to day 7 for twice-daily application over 4 days (22).

11.2.8 Nutmeg abuse and fatality

Nutmeg seeds are used as a spice and nutmeg oil has many benefits, including its antibacterial effect. Historically it was used as a stimulant and abortifacient, as well as for promoting menstruation. Nutmeg contains a volatile oil comprising several active compounds including myristicin. Nutmeg is abused in large quantities for its hallucinogenic effects because myristicin is metabolized to an active compound with an LSD-like effect. Owing to its euphoric and hallucinogenic effects, nutmeg has long been used as a low-cost substitute for other drugs of abuse. To obtain this effect, consumption of 15–20 g of nutmeg is needed, which can lead to severe toxicity. Stein et al. described a fatal case in which a 55 year-old woman died from nutmeg abuse as confirmed by the presence of myristicin in post mortem blood (23). Toxicity may occur on ingestion of approximately 5 g of nutmeg, corresponding to 1–2 mg of myristicin, but it is very unlikely that intake of nutmeg as a spice would cause any toxicity because the amount of myristicin and other active compounds ingested would be very low (24).

11.3 Other toxic herbal supplements

Several other herbal supplements, including pokeweed, skullcap, bitter orange and yohimbe, can also cause severe toxicity and should be avoided (▶Tab. 11.2). This section presents a brief discussion of each of these supplements.

11.3.1 Pokeweed poisoning

Pokeweed (American nightshade) is a large perennial herb that reaches a height of 10–12 ft (3.0–3.7 m) during the summer months and is found in Eastern parts of North America, California, Hawaii, Canada and other parts of the world. The berries and dried roots are used to prepare herbal remedies and have traditionally been used by Native American Indians for treating a variety of conditions, including skin disease, syphilis, cancer and infections, and as an emetic and narcotic. Unfortunately, all parts of pokeweed are toxic. Ingestion of uncooked berries also causes pokeweed poisoning. Although cooked young leaves (poke salad made by cooking the leaves twice and discarding the water) and cooked berries (after cooking twice and discarding the water) are eaten by some people, there is no guarantee that such cooked foods are safe for human consumption. Pokeweed toxicity increases with its maturity, but green berries are more toxic than red berries. Several cases of pokeweed poisoning have been reported in the literature. Thus, it is advisable to avoid pokeroot tea and herbal supplements containing pokeweed.

Tab. 11.2: Other herbal supplements with known toxicity

Bitter orange	Lobelia	Pokeweed	Skullcap	Guarana
Yohimbe	Mistletoe	Coltsfoot	Calamus	Kombucha mushrooms

Other herbs may also have adverse effects in humans. This list includes commonly encountered relatively toxic herbs but is not complete.

Despite its toxicity, pokeweed is used in certain herbal supplements to treat a number of conditions, including rheumatoid arthritis, tonsillitis, mumps, swollen glands, chronic excess mucus, bronchitis, mastitis, and constipation, as well as fungal infections, joint inflammation, hemorrhoids, breast abscesses, ulcers, and bad breath. Herbalists also claim that external application of a pokeweed preparation relieves itching, inflammation and skin diseases.

11.3.2 Skullcap toxicity

Skullcap (*Scutellaria lateriflora*) is a native North American plant that has been used for centuries as a folk remedy for treating anxiety, nervous tension and convulsion. It was once considered a remedy for treating rabies, but there is no scientific basis for this claim. Mainly the leaves are used for preparing herbal remedies. Chinese skullcap (*Scutellaria baicalensis*) has many pharmacological properties, including anti-inflammatory and anti-allergic actions. Despite its benefits, this herbal remedy should probably not be used owing to several reports of toxicity, especially liver toxicity. In one report, four women between 41 and 57 years of age experienced severe jaundice after taking a skullcap-containing herbal remedy to relieve stress. On discontinuation of the herbal supplement, their liver function tests returned to normal in 2–19 months (25). There are other reports of liver toxicity due to skullcap use in the literature.

11.3.3 Lobelia

Lobelia (*Lobelia inflata*), also called Indian tobacco, has been traditionally used for treating asthma and bronchitis. Lobelia leaves and seeds are used in making herbal remedies. Crude extract of lobelia may have antidepressant activity, but lobelia is a potentially toxic herb and ingestion may cause nausea, vomiting, rapid heartbeat, low blood pressure and possibly coma. Therefore, it is advisable to avoid lobelia-containing herbal supplements.

11.3.4 Androstenedione

Androstenedione is a steroid that is produced naturally in the body and it is a common precursor of the male sex hormone testosterone and the female sex hormones estrogen, estradiol and estrone. Androstenedione was manufactured and sold as a dietary supplement called Andro to improve sports performance and as an aid for body building and weight loss. One survey in 2002 revealed that 1 out of 40 high school students had used Andro in the previous year. The FDA banned the sale of this dietary supplement in 2004, citing health risks involved in Andro use, and sent letters to 23 companies asking them to stop distributing products sold as dietary supplements that contain androstenedione and warning them that they could face enforcement action if they did not take appropriate action (26).

11.3.5 Bitter orange

Bitter orange or Seville orange has been used in traditional Chinese medicine and today is used mainly as a weight loss product and a nasal decongestant. It is also used in treating indigestion and nausea. Bitter orange is used topically to treat ringworm and athlete's foot. Following withdrawal of ephedrine from the dietary supplement market, weight loss products containing bitter orange have been increasing in popularity. At this point there is little evidence that bitter orange may promote weight loss. However, synephrine and other structurally related compounds are present in bitter orange and these compounds are structurally similar to ephedra. Ingestion of these herbal supplements may increase blood pressure and heart rate. Between January 1, 1998 and February 28, 2004, Health Canada received 16 reports for which products containing bitter orange or synephrine were suspected of being associated with adverse cardiovascular effects such as blackout, transient collapse, cardiac arrest, tachycardia and ventricular fibrillation (27). People with heart disease or high blood pressure should not take products containing bitter orange as it may cause resistant hypertension. A 55-year-old man who took 300 mg of bitter orange extract every day had an acute myocardial infarction (28). Bitter orange oil when applied to skin may cause photosensitive reactions.

11.3.6 Guarana

Guarana, a popular herbal weight loss product (also used as an ingredient in many herbal weight loss products) is prepared from seeds of guarana and contains 2.5–7% caffeine (200 mg/dose). By contrast, one cup of coffee contains 100 mg or less of caffeine. At the guarana dose recommended for weight loss, a person may take up to 1800 mg of caffeine per day. Such a high caffeine intake may lead to adverse effects such as an increase in blood pressure to a dangerous level in a person already suffering from high blood pressure. If a person taking such a supplement also drinks coffee or takes a medicine such as pseudoephedrine that may also increase blood pressure, there may be an additive effect on blood pressure and cardiac problems may arise. A 25-year-old woman with preexisting mitral valve prolapse died after using a herbal supplement containing guarana and ginseng (see color plate) owing to the high caffeine content of the preparation (29).

11.3.7 Yohimbe

Yohimbe bark has been traditionally used in Africa to increase sexual desire and is now used as a dietary supplement for treating sexual dysfunction, including erectile dysfunction in men. Yohimbe bark can be brewed as a tea and consumed. A bark extract is also available commercially. Although yohimbe is effective in treating erectile dysfunction, its benefits seem to be outweighed by its risks (30). Yohimbe use has been associated with high blood pressure, increased heart rate, dizziness and other symptoms and if taken in large doses for a long time can be dangerous. A 43-year-old man died in a hotel room during a sexual relation with a colleague. He was taking Viagra and pill boxes found in the room contained several other drugs, including yohimbe (31). Therefore, yohimbe does not have a good safety record and should probably be avoided.

11.3.8 Other toxic herbal supplements

Young flower buds of coltsfoot are used in herbal remedies and contain senkirkine, a hepatotoxic pyrrolizidine alkaloid that has carcinogenic activity (32). Therefore, herbal supplements containing coltsfoot must be avoided. *Acorus calamus* (calamus or sweet flag) preparations are available via the Internet and are marketed as a hallucinogenic herbal supplement. From 2003 to 2006, the Swedish Poison Control Center received inquiries about 30 clinical cases of intentional intoxication induced by calamus-containing herbal supplements. The main symptom of calamus intoxication is prolonged vomiting, which may last for up to 15 h. α-Asarone and β-asarone can be detected in urine from patients intoxicated with calamus, in addition to another metabolite, 2,4,5-trimethoxyamphetamine (33).

Herbalists claim that Kombucha mushroom tea has medicinal properties, but tea prepared from this mushroom is toxic. Kombucha tea is a black tea fermented in a yeast/bacteria medium. Sung Hee Kole et al. reported a case of a 22-year-old male newly diagnosed with HIV who became short of breath and febrile with hyperthermia and presented to hospital with lactic acidosis and acute renal failure within 15 h after consuming Kombucha tea. He subsequently became combative and confused, requiring sedation and intubation for airway control. He had elevated serum lactate of 12.9 mmol/L and serum creatinine of 2.1 mg/dL (187 µmol/L). Kombucha tea can cause serious hepatic damage and even fatality (34).

11.4 Conclusion

Contrary to the popular belief that herbal supplements are natural and therefore safe, there are many reports of severe toxicity and even death from consumption of certain toxic herbal supplements. As more research results accumulate in the field of herbal supplements, a better understanding of the safety of herbal supplements can be obtained. In this chapter, commonly used herbal supplements that are toxic have been discussed. The list is not complete, as some lesser-known herbal supplements may also cause toxicity and a herbal supplement considered to be relatively safe may be reclassified as moderately toxic or toxic as a result of future investigations. A good example is Kava, which was considered as safe and effective prior to 1998 but it is now well documented that this herbal supplement may cause significant liver damage and even death. Clinicians and healthcare professionals should be aware of such toxic herbal products so that they can warn their patients to avoid critical or toxic effects.

References

1. Lazarou, J., Pomeranz, B.H., Corey, P.N. (1998) Incidence of adverse drug reactions in hospitalized patients: a meta-analysis of prospective studies. J. Am. Med. Assoc. 279:1200–1205.
2. Dennehy, C.E., Tsourounis, C., Horn, A.J. (2005) Dietary supplement related adverse events reported to the California Poison Control System. Am. J. Health Syst. Pharm. 62:1476–1482.
3. Wang, Z., Zhao, J., Xing, J., He, Y., Guo, D. (2004) Analysis of strychnine and brucine in post mortem specimens by RP-HPLC: a case report of fatal intoxication. J. Anal. Toxicol. 28:141–144.
4. Zhu, S.H., Ananda, S., Yuan, R.X., Ren, L., Chen, X.R., Liu, L. (2010) Fatal renal failure due to the Chinese herb GuanMu Tong (Aristolochia manshuriensis): autopsy findings and review. Forensic Sci. Int. 199:e5–7.
5. Haller, C.A., Benowitz, N.L. (2000) Adverse and central nervous system events associated with dietary supplements containing ephedra alkaloids. N. Engl. J. Med. 343:1833–1838.

6. Chen-Scarabelli, C., Hughes, S.E., Landon, G., Rowley, P., Allebban, Z., Lawson, N. et al. (2005) A case of fatal ephedra intake associated with lipofuscin accumulation, caspase activation and cleavage of myofibrillary proteins. Eur. J. Heart Fail. 7:927–930.

7. Naik, S.D., Freudenberger, R.S. (2004) Ephedra-associated cardiomyopathy. Ann. Pharmacother. 38:400–403.

8. Haynes, B.E., Bessen, H.A., Wightman, W.D. (1985) Oleander tea: herbal draught of death. Ann. Emerg. Med. 14:350–353.

9. Eddleston, M., Senarathna, L., Mohammed, F., Buckley, N., Juszczak, E., Sheriff, M.H. et al. (2000) Deaths due to the absence of an affordable antitoxin for plant poisoning. Lancet 362:1041–1044.

10. Ko, R., Greenwald, M., Loscutoff, S., Au, A., Appel, B.R., Kreutzer, R.A. et al. (1996) Lethal ingestion of Chinese tea containing Chan Su. Western J. Med. 164:71–75.

11. Gordon, D.W., Rosenthal, G., Hart, J., Sirota, R., Baker, A.L. (1995) Chaparral ingestion: the broadening spectrum of liver injury caused by herbal medications. J. Am. Med. Assoc. 273:489–490.

12. Gow, P.J., Connelly, N.J., Hill, R.L., Crowley, P., Angus, P.W. (2003) Fatal fulminant hepatic failure induced by a natural therapy containing kava. Med. J. Aust. 178:442–443.

13. Mostefa-Kara, N., Pauwels, A., Pines, E., Biour, M., Levy, V.G. (1992) Fatal hepatitis after herbal tea. Lancet 340:674.

14. Chou, W.C., Wu, C.C., Yang, P.C., Lee, Y.T. (1995) Hypovolemic shock and mortality after ingestion of Tripterygium wilfordii Hook F: a case report. Int. J. Cardiol. 49:173–177.

15. Salminen, A., Lehtonen, M., Paimela, T., Kaarniranta, K. (2010) Celastrol: molecular targets of thunder god vine. Biochem. Biophys. Res. Commun. 394:439–442.

16. But, P.P., Tai, Y.T., Young, K. (1994) Three fatal cases of herbal aconite poisoning. Vet. Hum. Toxicol. 36:212–215.

17. Pullela, R., Young, L., Gallagher, B., Avis, S.P., Randell, E.W. (2008) A case of fatal aconitine poisoning by monkshood ingestion. J. Forensic Sci. 53:491–494.

18. Cahn, T.Y. (2009) Aconite poisoning. Clin. Toxicol. 47:279–285.

19. Davis, J.E. (2007) Are one or two dangerous? Methyl salicylate exposure in toddlers. J. Emerg. Med. 32:63–69.

20. Chan, T.Y. (2002) Ingestion of medicated oils by adults: the risk of severe salicylate poisoning is related to the packaging of these products. Hum. Exp. Toxicol. 21:171–174.

21. Parker, D., Martinez, C., Stanley, C., Simmons, J., McIntyre, I.M. (2004) The analysis of methyl salicylate and salicylic acid from Chinese herbal medicine ingestion. J. Anal. Toxicol. 28:214–216.

22. Morra, P., Bartle, W.R., Walker, S.E., Lee, S.N., Bowles, S.K., Reeves, R.A. (1996) Serum concentrations of salicylic acid following topically applied salicylate derivatives. Ann. Pharmacother. 30:935–940.

23. Stein, U., Greyer, H., Hentschel, H. (2001) Nutmeg (myristicin) poisoning – report on a fatal case and a series of cases recorded by a poison information center. Forensic Sci. Int. 118:87–90.

24. Hallstrom, H., Thuvander, A. (1997) Toxicological evaluation of myristicin. Nat. Toxin 5:186–192.

25. MacGregor, R.F.B., Abernethy, V.E., Dahabra, S., Cobden, I., Hayes P.C. (1989) Hepatotoxicity from herbal remedies. Br. Med. J. 299:1156–1157.

26. Anonymous (2004) Crackdown on 'andro' products. FDA Consumer 38:26.

27. Haaz, S., Fontaine, K.B., Cutter, G., Limdi, N., Perumean-Chaney, S., Allison, D.B. (2006) Citrus aurantium and synephrine alkaloids in the treatment of overweight and obesity: an update. Obes. Res. 7:79–88.

28. Hess, A.M., Sullivan, D.L. (2005) Potential for toxicity with use of bitter orange extract and guarana for weight loss. Ann. Pharmacother. 39:574–575.

29. Cannon et al. Caffeine-induced cardiac arrhythmia: an unrecognised danger of healthfood products. Med. J. Aust. 2001;174:520–521.
30. Ernst, E., Pittler, M.H. (1998) Yohimbine for erectile dysfunction: a systematic review and meta-analysis of randomized clinical trials. J. Urol. 159:433–436.
31. Dumestre-Toulet, V., Cirimele, V., Gromb, S., Belooussoff, T., Lavault, D., Ludes, B. et al. (2002) Last performance with Viagra: post-mortem identification of sildenafil and its metabolites in biological specimens including hair sample. Forensic Sci. Int. 126:71–76.
32. Hirono, I., More, H., Culvenor, C.C. (1976) Carcinogenic activity of coltsfoot, Tussilago farfara. Gann 67:125–129.
33. Bjornstad, K., Helander, A., Hulten, P., Beck, O. (2009) Bioanalytical investigation of asarone in connection with Acorus calamus oil intoxications. J. Anal. Toxicol. 33:604–609.
34. SungHee Kole, A., Jones, H.D., Christensen, R., Gladstein, J. (2009) A case of Kombucha tea toxicity. J. Intensive Care Med. 24:205–207.

Appendix

Tab. A1: Abnormal clinical laboratory test results due to use of herbal supplements

Herbal supplement	Abnormal test result	Comments
Abnormal liver function tests		
Kava	Highly elevated liver enzymes	Kavalactones cause liver damage
		Death attributed to kava toxicity
Chaparral, comfrey	Significantly elevated liver enzymes	Liver toxicity is a major concern
Germander	Significantly elevated liver enzymes	Teucrin A is a liver toxin
LipoKinetix	May elevate liver enzymes	Sodium usniate causes liver damage
Pennyroyal	May elevate liver enzymes	Pulegone is a liver toxin
Mistletoe	May elevate liver enzymes	Viscotoxins and lectins are liver toxins
Abnormal kidney function tests		
Chinese weight loss products	Elevated creatinine and urea	Certain Chinese weight loss products containing aristolochic acid
Calamus, sassafras, wormwood	Elevated creatinine	These herbs may damage kidney
Abnormal thyroid profile		
Kelp	Abnormal TSH, T3 and T4	Kelp contains high amounts of iodine and interferes with thyroid function
Unexpected hypoglycemia		
Ginseng, fenugreek	Lower glucose level	May lower glucose but not suitable for treating diabetes
Chromium	Lower glucose level	Chromium can sensitize insulin, causing hypoglycemia
Milk thistle	May lower glucose	
Unexpectedly elevated digoxin		
Oleander, Chan Su, Lu-Shen-Wan	Significantly elevated digoxin	Supplements interfere with digoxin immunoassays

Tab. A2: Clinically significant drug-herb interactions

Herbal supplement	Interacting drug	Comments
Boldo	Warfarin	Risk of bleeding
Borage	Warfarin	Risk of bleeding
Bromelain	Naproxen	Increased bleeding
Coenzyme Q10	Hydrochlorothiazide Metformin, Glipizide Warfarin	Hypotension Hypoglycemia Risk of bleeding
Devil's claw	Warfarin	Risk of bleeding
Echinacea	Ketoconazole, methotrexate Simvastatin, lansoprazole	Liver toxicity Increased effect
Evening primrose oil	Warfarin	Risk of bleeding
Fenugreek	Warfarin	Risk of bleeding
Feverfew	Iron Cyclosporine, azathioprine Tacrolimus Warfarin	Reduced absorption Reduced effect Reduced effect Risk of bleeding
Fish oil	Metoprolol Warfarin	Hypotension Risk of bleeding
Flaxseed oil	Aspirin	Increased bleeding
Garlic	Saquinavir Ibuprofen Chlorpropamide Warfarin	Reduced effect Bleeding Hypoglycemia Risk of bleeding
Ginkgo biloba	Aspirin, clopidogrel Dipyridamole, ticlopidine Trazodone Phenytoin, valproic acid Warfarin	Bleeding Bleeding Coma Reduced effects Risk of bleeding
Ginger	Warfarin	Risk of bleeding
Glucosamine	Acetaminophen Glyburide	Reduced effect Hyperglycemia
Golden seal	Warfarin	Reduced efficacy
Green tea	Warfarin	Reduced efficacy
Horse chestnut	Warfarin	Risk of bleeding
Kava	Alprazolam, benzodiazepines	Coma
Licorice	Atenolol	Hypotension
Milk thistle	Warfarin	Reduced efficacy
Saw palmetto	Aspirin Warfarin	Bleeding Risk of bleeding

(Continued)

(*Continued*)

Herbal supplement	Interacting drug	Comments
St. John's wort	Cyclosporine, tacrolimus	Reduced efficacy
	Warfarin	Reduced efficacy
	Indinavir, saquinavir, atazanavir	Reduced efficacy
	Lamivudine, nevirapine	Reduced efficacy
	Imatinib, irinotecan	Reduced efficacy
	Digoxin, verapamil, nifedipine	Reduced efficacy
	Alprazolam, midazolam, quazepam	Reduced efficacy
	Mephenytoin, phenobarbital	Reduced efficacy
	Gliclazide	Reduced efficacy
	Erythromycin, voriconazole	Reduced efficacy
	Theophylline	Reduced efficacy
	Omeprazole	Reduced efficacy
	Simvastatin, atorvastatin	Reduced efficacy
	Ethinylestradiol and related compounds	Reduced efficacy
	Amitriptyline, methadone, oxycodone	Reduced efficacy
Valerian	Midazolam, barbiturates	Increased sedation

Tab. A3: Important fruit juice-drug interactions

Fruit juice	Interacting drug	Comments
Grapefruit	Felodipine	Several fold increase in blood level
	Nitrendipine	Increased bioavailability by 100%
	Pranidipine	73% increase in bioavailability
	Nimodipine	Modest increase in bioavailability
	Amiodarone	Significantly increased AUC
	Simvastatin	AUC may be increased 3.6 fold
	Atorvastatin	AUC increased by 83%
	Pitavastatin, lovastatin	Modest increase in AUC
	Diazepam, triazolam, midazolam	Significantly increased blood level
	Carbamazepine	Increased blood level
	Cyclosporine, tacrolimus	Modest to significant increase in blood levels
	Saquinavir	Significant increase in blood level
	Itraconazole	Modest increase in blood level
	Colchicine	Increased toxicity
	Fexofenadine	Reduction in blood level
	Digoxin	No interaction
	Phenytoin	No interaction
Orange	Pravastatin	Significant increase in AUC
	Celiprolol	AUC reduced by 83%
Seville orange	Felodipine	AUC increased by 76%
Pomelo	Cyclosporine, tacrolimus	Increased blood level
	Sildenafil	Reduced blood level
Pomegranate	Carbamazepine	Increased AUC by 1.5 fold
	Warfarin	Increased effectiveness
Cranberry	Warfarin	INR >50 in a patient causing death

Index

Color plates

Echinacea cone flower

Ginkgo leaves

Garlic bulbs

Ginseng

Ginger roots

St. John's wort